SIZING UP THE
UNIVERSE

SIZING UP THE
UNIVERSE

THE COSMOS IN PERSPECTIVE

J. Richard Gott / Robert J. Vanderbei

NATIONAL GEOGRAPHIC

WASHINGTON, D.C.

As Jupiter's moon Io transits in front of Jupiter, it can be seen against the upper layer of Jupiter's dense clouds below.

■CONTENTS

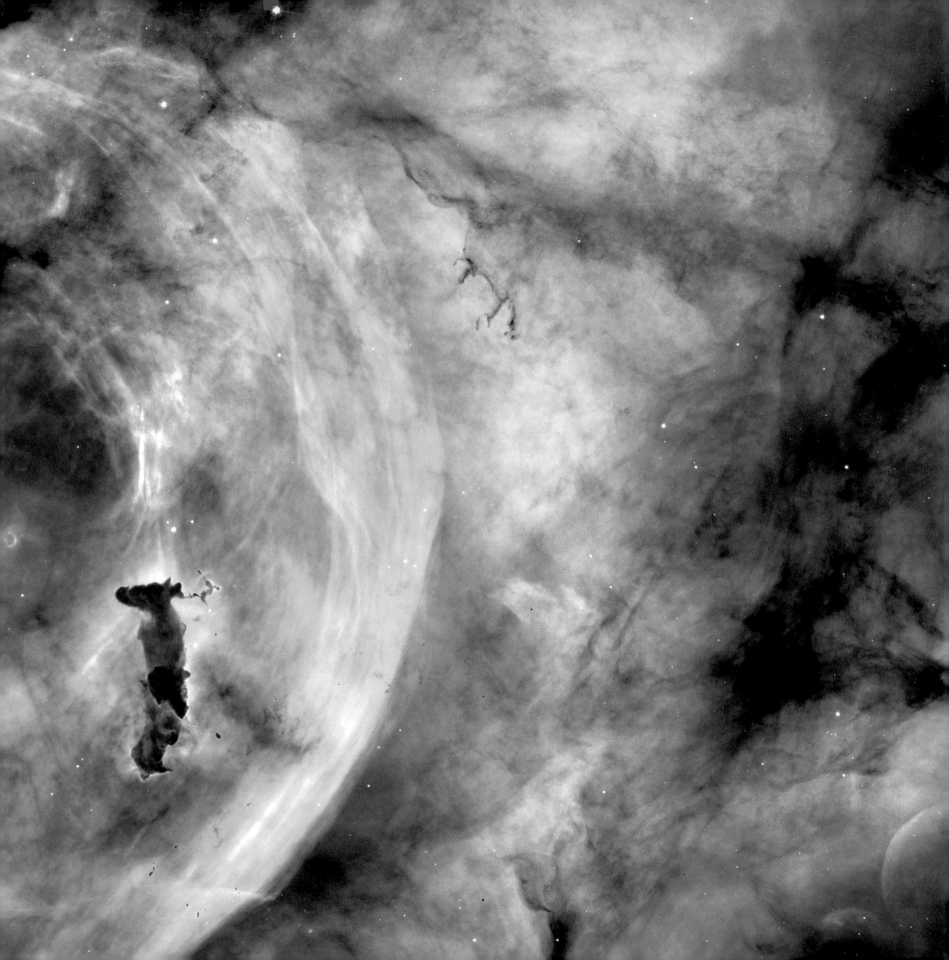

In this close-up view, the Carina Nebula features a pair of Bok globules (left and right), which are thought to be concentrations of dust condensing into yet-to-be-born stars.

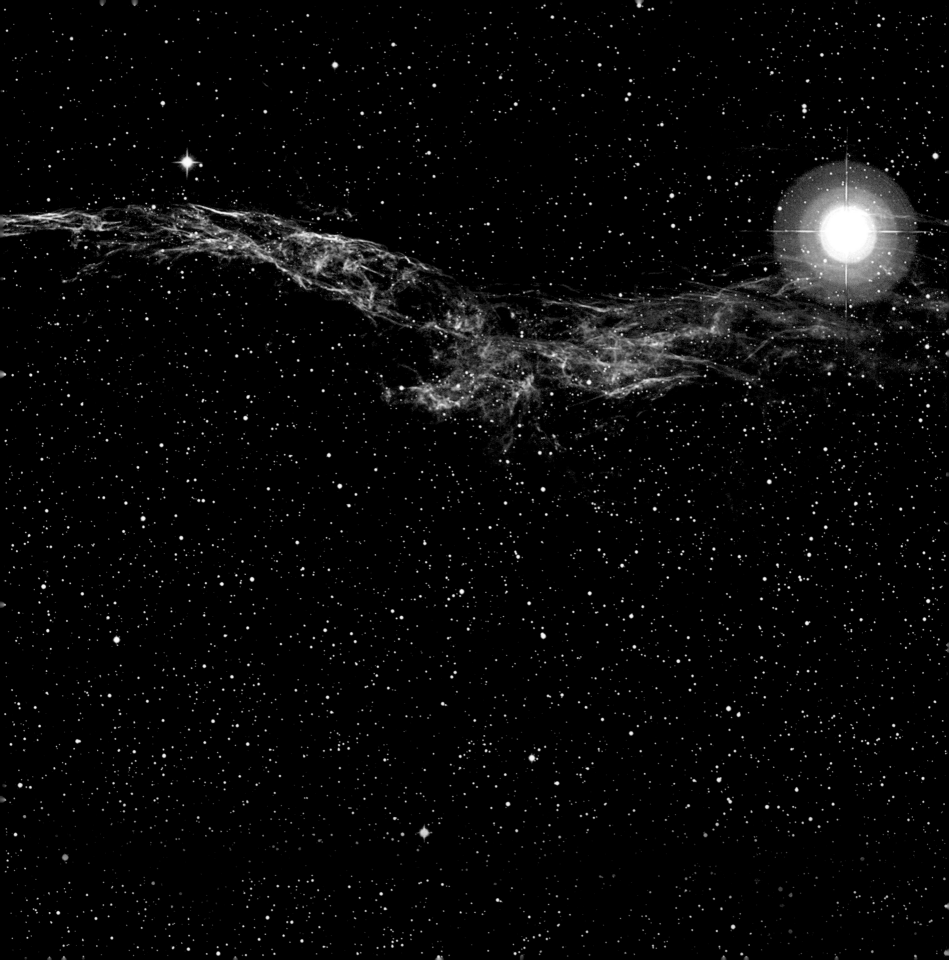

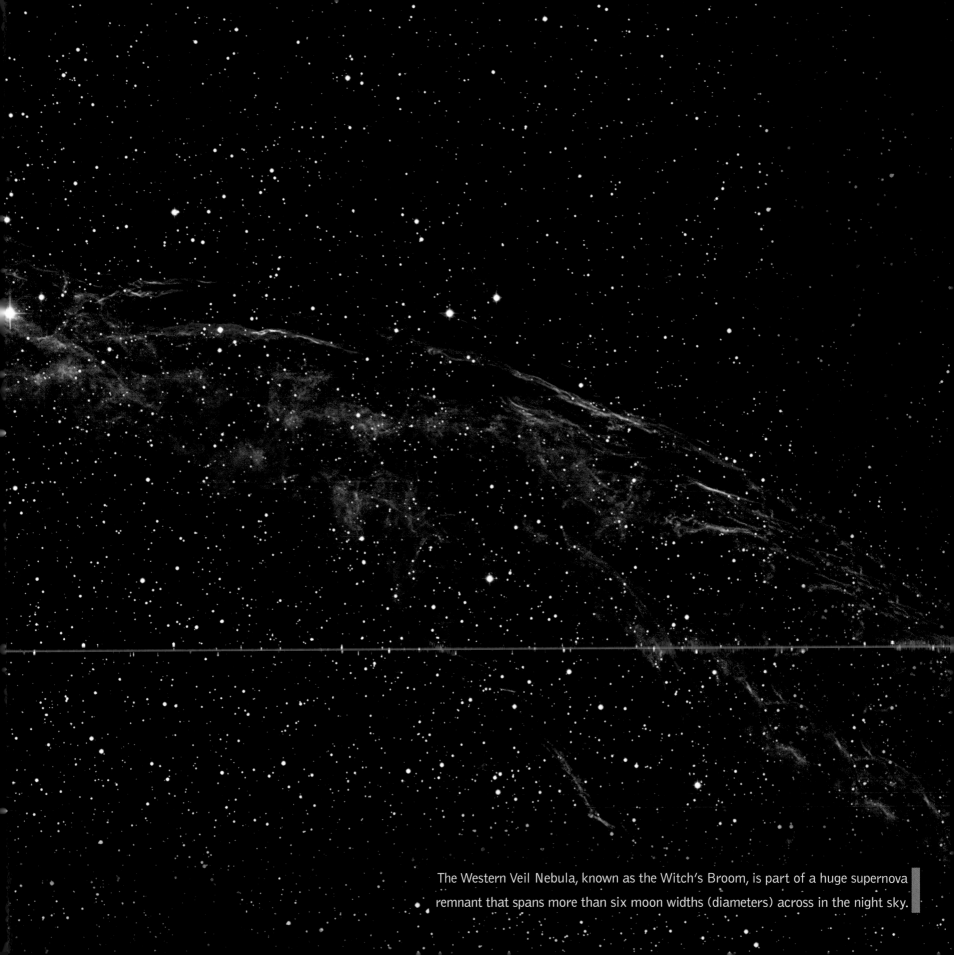

The Western Veil Nebula, known as the Witch's Broom, is part of a huge supernova remnant that spans more than six moon widths (diameters) across in the night sky.

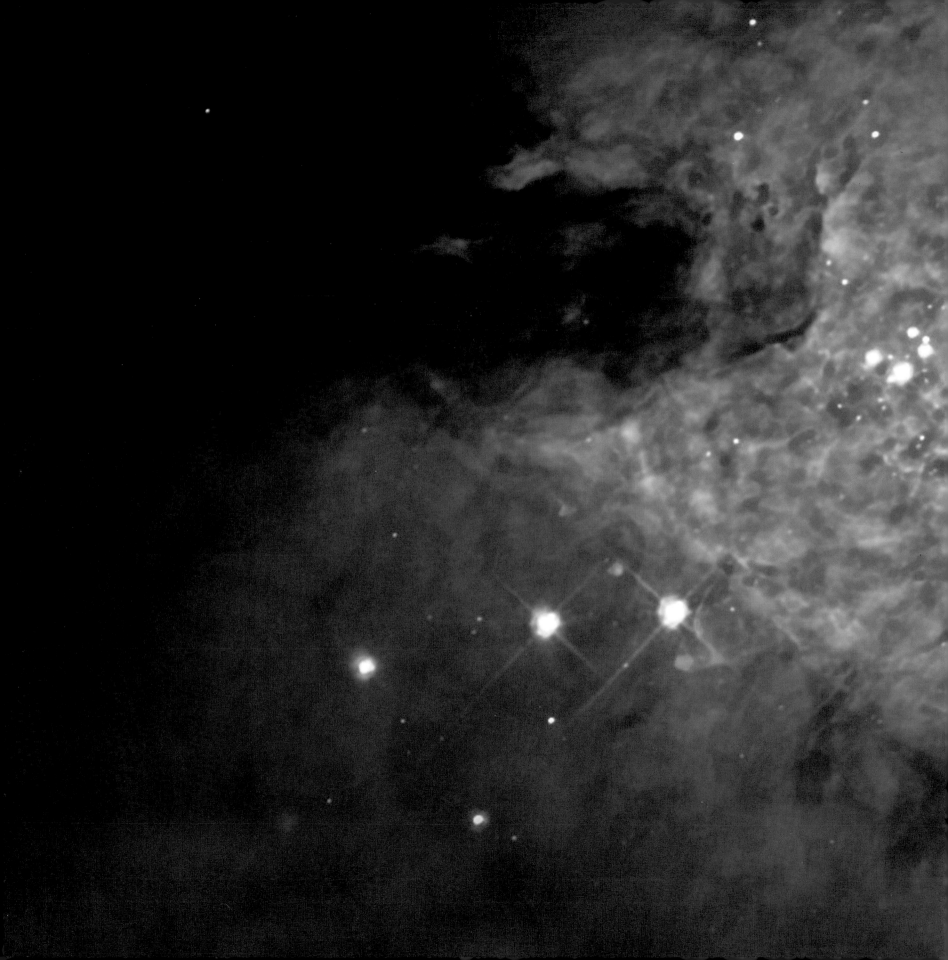

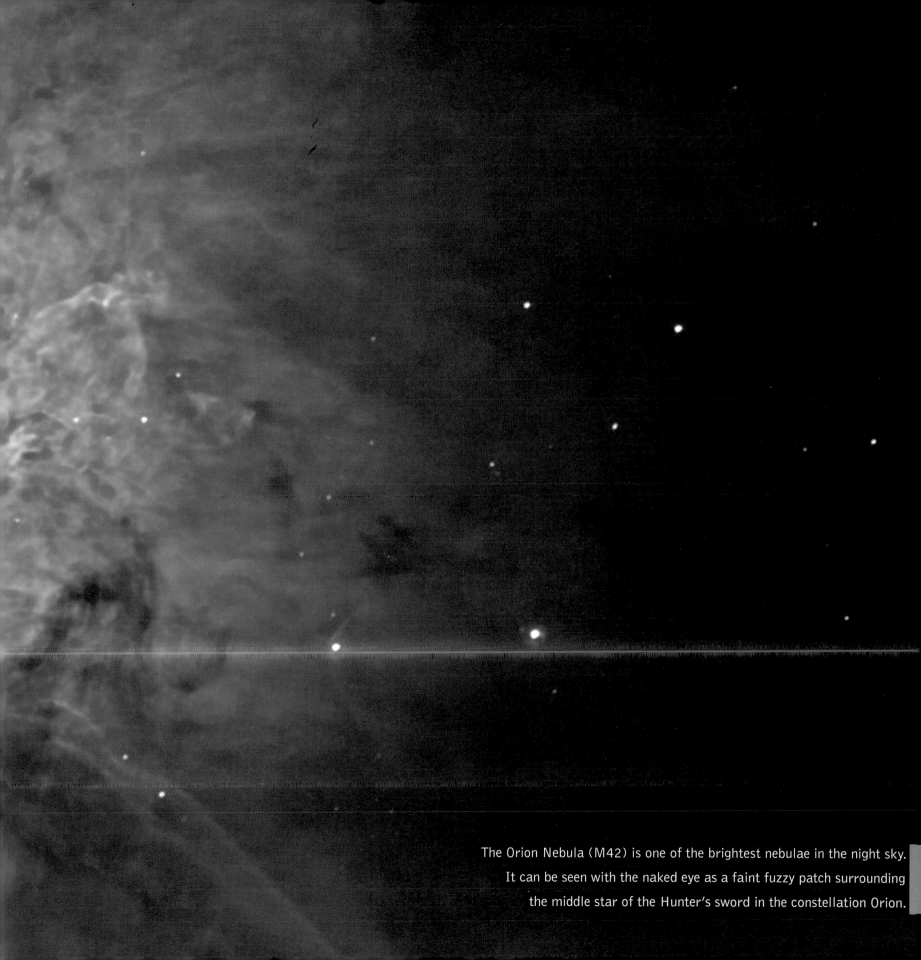

The Orion Nebula (M42) is one of the brightest nebulae in the night sky. It can be seen with the naked eye as a faint fuzzy patch surrounding the middle star of the Hunter's sword in the constellation Orion.

PREFACE

I BECAME INTERESTED in astronomy when I was eight years old. My mother took me to a bookstore in town, and there I found a book called *Stars* by Herbert Zim and Robert Baker. I still have that well-worn copy today. What I found most intriguing were the sizes of the planets and the sun. I learned them all. Jupiter had a diameter of more than 88,000 miles, and the sun had a diameter of 860,000 miles. I still remember those numbers today. I think this is one reason little kids like astronomy: Astronomy deals with things that are huge. Dinosaurs are also popular with kids because they are so large. But planets and stars are way bigger than dinosaurs—so I went into astronomy!

I have been interested in visualizing the universe for a long time. As a teenager, I thought a map of the celestial sphere projected onto a cube would be particularly elegant, since (as I explain in Chapter 2) it would plot great circles as straight lines and would produce six star charts that conveniently divide the celestial sphere into the north circumpolar, south circumpolar, summer, autumn, winter, and spring stars. I made this my eighth-grade science-fair project (see picture opposite). It won first place in the Junior Discussion Division of the Kentucky State Science Fair.

Years later, I happened on a book in Princeton's astrophysics library: *Six Maps of the Stars,* published in 1831 by Baldwin and Cradock, with charts by W. Newton. It was exactly the same idea. The book said similar charts had been made going back as far as 1674!

In graduate school, I became interested in producing a map of the entire visible universe—one grand map that would show everything from satellites orbiting Earth to planets, stars, and galaxies out to the most distant things we can see. A version of this map that I created with Mario Jurić was published in 2003 and has been

reprinted 1.5 million times, in the *New York Times* and in *New Scientist* and *Astronomy* magazines and elsewhere. (See pages 121-128 for an updated version of the map.)

To make our Map of the Universe, we used data from the Sloan Digital Sky Survey, a project that mapped a large part of the sky. In the process of making our map, Mario Jurić and I measured a long, dense aggregation of galaxies, brought together by gravity. We named it the Sloan Great Wall of galaxies and found it to be the largest structure in the universe. It has a length of 1.37 billion light-years. That means it would take light, traveling at 186,000 miles per second, 1.37 billion years to travel from one end of the wall to the other.

Jurić and I later found ourselves and the Sloan Great Wall of galaxies in the book *Guinness World Records 2006.* This book has many "largest" things—the largest pickup truck or the largest tree, for example. But of all the "largest" things in the book, our Sloan Great Wall of galaxies is the largest of them all. I was quite surprised to find myself in the *Guinness World Records,* and I didn't even have to eat an enormous number of hot dogs or collect the world's largest ball of twine!

That adventure reaffirmed my lifelong interest in the sizes of things in the universe. Today I am probably best known for my research on cosmology and the theory of general relativity, especially its implications for time travel. That research is described in my book *Time Travel in Einstein's Universe.* Working on the Map of the Universe and the Sloan Great Wall encouraged me to write a book portraying sizes in the universe. I just needed a collaborator to help me produce the illustrations I envisioned. I found one in Bob Vanderbei. He is chair of the Department of Operations Research and Financial Engineering at Princeton, and an outstanding leader in his field. He is also an associated faculty member of Princeton's Department of Astrophysical Sciences.

Vanderbei is perhaps most famous for his Purple America map, showing how the votes divided county by county in the presidential elections starting in 2000. Before that, it was often said that the United States was divided into red states that voted Republican and blue states that voted Democratic. Vanderbei colored in each county with red proportional to the Republican vote, blue proportional to the Democratic vote, and green proportional to the vote for the Independent candidate Ralph Nader. The result was various shades of purple, a mixture of red and blue. As it turned out, the country was less divided than we thought. Vanderbei's map was reprinted in *Time, Newsweek,* and elsewhere and often appeared on TV. So here was an illustrator par excellence.

Vanderbei is also an amateur astronomer. His photographs of astronomical objects taken right from his backyard rival the best images taken from observatories. He became interested in astronomy later in life, after attending a "star party" in 1998, where amateurs showed

off the wonders of the sky through their telescopes. Vanderbei was particularly captivated by the view of Jupiter. Not only could you see cloud belts and moons orbiting it, but at one point in the evening, the shadow of Io, one of Jupiter's large moons, began to pass across the disk of the planet. Vanderbei was hooked. He had been a glider pilot, but he sold his plane and bought a telescope. Vanderbei and I came to astronomy at different times. I started as a kid, whereas Vanderbei took up this hobby as an adult—yet we are both typical. When I went to my first national convention of amateur astronomers in the 1960s, as a teenager, about 20 percent of the people at the convention were under 18. At a similar convention of the Astronomical League in recent years, only 3 percent of the participants were under 18. So what happened?

Richard Gott's eighth-grade science-fair project (above). Vanderbei's photo of Jupiter's Great Red Spot (opposite) shows the shadows of two of Jupiter's moons, Io and Europa.

Years ago, when I started astronomy, many young people were discovering it as a backyard hobby. You could see the wonders of the universe—from planets to galaxies—all from your backyard. But our increasingly bright city lights over the years have brightened the sky so much that the galaxies have faded from view. We call this phenomenon light pollution. To see anything but the brightest objects in the night sky you now have to go to a dark site. The backyard hobby became one where you needed to drive to a remote location and stay all night. As a result, it became a hobby for adults. Like skiing, it had cool equipment, but while better skis might make you only a slightly better skier, a bigger and more expensive telescope would allow you to see a lot more. So Vanderbei caught the wave of adults becoming amateur astronomers.

Vanderbei is also part of the making of a new hobby—astrophotography. The new digital cameras available today allow you to take long-exposure photographs and use filters to overcome light pollution. You can take photographs from your backyard that will show things you would never be able to see through the telescope. The galaxies are back! For the moon and planets, a simple webcam attached to a small telescope allows you to take videos of these objects and select the best frames (those least affected by atmospheric distortion) for producing your picture.

Vanderbei has taken all of his photographs from his backyard in light-polluted New Jersey. He took a beautiful picture of the lunar crater Plato using only a 3.5-inch-diameter telescope and a simple webcam. His photographs appear throughout this book along with the best NASA photographs, as noted in the illustrations credits.

I have had the good fortune to travel around the world and see a number of its wonders, from the Taj Mahal to the Great Pyramid, and one of my joys was taking pictures along the way. But Vanderbei can take pictures of the wonders of the universe without ever leaving home. Vanderbei sets his telescope tracking a galaxy for an hour exposure and goes inside to watch television while it takes the picture. He can then download the picture to his computer and sharpen its image with software or combine it with other exposures to make a brighter image.

Astronomy has become a computer hobby. Young people today are experts at downloading digital photographs onto their computers. We hope that this book, by showing what can be done from the backyard with a digital camera and a telescope, will help lure young astronomers back, for it has once again become a backyard hobby. With that in mind, Vanderbei has written an afterword with some of his thoughts on astrophotography, hoping that some of our readers will be encouraged to size up the universe for themselves.

Whether this is your first book on astronomy or you have read many, there should be some surprises here. (For example, I myself was surprised to see just how big the Ring Nebula is—I had thought it was about 25 times smaller.) Whether you are 16 years old or 60, whether you enjoy looking through a telescope or are just a curious armchair astronomer who has heard of black holes and wants to know more, this book was made for you. I hope you will find learning about sizes in the universe just as fascinating as I did once, so long ago.

—*J. Richard Gott*

Vanderbei captured this photo of the Eagle Nebula (opposite) with a 10-inch-diameter telescope from his backyard. For comparison, the famous Hubble Space Telescope picture of the Eagle Nebula is shown inset. The Hubble picture has more detail, having been taken from the vacuum of space, but the colors in Vanderbei's photograph are more realistic—the nebula is actually red.

INTRODUCTION

HOW BIG ARE THINGS in the sky? It's a child's question, really, but one that has altered the course of civilization. When Aristarchus of Samos first figured out that the sun was bigger than Earth, in about 260 B.C., he correctly deduced that Earth must be orbiting the larger sun, rather than the sun circling tiny Earth. But people didn't believe him. They followed Aristotle's idea that Earth was at the center of the universe. It wasn't until Copernicus published the same idea in 1543 that the concept began to be taken seriously. If people had believed Aristarchus in the first place, science would have been pushed forward by more than 1,700 years. Where might our civilization and technology be by now if people had just listened?

One of the exciting things about astronomy is just *how big* those things in the sky really are. The sun is so big that a million Earths could fit inside. And yet there are stars much larger than the sun. The distances between the stars and between the galaxies are truly humbling. Our tiny Earth is just a speck in the vastness of space. It tends to put things into perspective.

In Chapter 1 we start by showing how big things *look* in the sky. The sun and the moon are the same apparent size in the sky—each about one-half degree across. As they set, you can cover each of them with your thumb held at arm's length. If you know how big something looks and how far away it is, you can figure out its real size. Multiplying an object's apparent size by its distance gives its actual size. Although the sun and the moon appear to be the same size in the sky, the sun is actually 400 times farther away, so its actual size is 400 times larger than the moon—a simple concept but an important one.

Astronomers typically learn how big an object looks by taking its picture through a telescope of known power.

Then when we learn how far away it is, we can determine its actual size. Thus, we begin with side-by-side pictures of the sun and moon, showing how big they look. Then we show a series of ever more magnified views of objects through ever more powerful telescopes. As we continue our observing tour, we will show you how big things would look through the Hubble Space Telescope.

Then, in Chapters 2-4, we show where things are in the sky and how far away they are, keyed to discoveries by the ancient thinkers who were not as far off in their calculations as one might expect. Some of their methods are still in use today. And, in Chapter 4, with the conceptual framework firmly established, we offer a "map" of the universe, showing everything from satellites in Earth's orbit to distant stars and galaxies, all on one fold-out section almost 40 inches tall.

Once we know how big things *look,* and how *far away* they are, we can show the *actual sizes* of things in Chapters 5 and 6. Chapter 5 shows sizes of things in the solar system. Hurricane Katrina is compared with a centuries-old storm on Jupiter called the Great Red Spot. In Chapter 6 we start with an actual-size picture of Buzz Aldrin's footprint on the moon and end with the largest known object in the cosmos. If you are someone who likes to peek at the last page of a mystery novel to find out the answer, and you want to know about the real sizes of things, you can skip directly to Chapters 5 and 6.

Photograph of the Horsehead Nebula taken by Bob Vanderbei with his 10-inch-diameter reflecting telescope. Glowing gas in the background is obscured by dust to make the horse's head.

How to Use This Book

THIS IS A BOOK about sizing up the universe, figuring out how big things are and how far away they are—mapping the universe and its contents.

Chapter 1 begins with the angular sizes of things in the sky—how big objects appear in the sky as seen from our earthbound perspective. In it, we offer visual comparisons that are signaled by an orange bar across the top of the pages showing the magnification of the view (seen from a reading distance of 17 inches). In most of these cases, the objects would never appear next to each other in the sky. We juxtapose them to compare them.

We start off with a naked-eye view, then show a series of ever more magnified views. On the opposite page, top, is a miniature version of the 23× magnification view. It shows the moon with the Andromeda galaxy in the background. The white line segment at the bottom shows the angular size—in this case, 17 minutes of arc. Labels indicate the distances of the objects from us. As we continue, we will show views with magnification equal to that achieved with the Hubble Space Telescope and beyond.

In Chapter 5, we show side-by-side pictures to scale of different objects in the solar system. Finally, in Chapter 6, we present a series of pictures, each covering a scale a thousand times larger than the one before, to survey the sizes of objects in the universe. The first picture shows Buzz Aldrin's footprint on the moon—actual size. The fourth picture in this sequence (shown opposite, middle) depicts objects in the solar system all at a scale of 1:1 billion. For each picture in this sequence, an orange bar at the top gives the scale. The white line segment at the bottom illustrates the physical size—in this case, 50,800 kilometers.

In Chapters 5 and 6 we offer many comparisons in size and scale in order to give the reader a new perspective on the cosmos. Sidebar sections, indicated by a gray bar across the top of the pages, use lively typography, diagrams, or both to elucidate key concepts.

Throughout we have scattered innovative graphics and dramatic stereoscopic images (shown opposite, bottom) to illustrate the wonders of the universe. Highlights from recent research on everything from the origin of Saturn's rings to the spongelike clustering pattern of galaxies to universes beyond our own are woven through the story.

We are presenting this new view of the cosmos at a time when recent discoveries have changed our cosmic perspective. For the first time, we now know precisely how big the visible universe actually is. From icy bodies in the outer solar system to new planets orbiting other stars to giant walls of galaxies, exciting discoveries big and small have illuminated the true diversity in the universe.

Helpful Conversions

ANGLES
There are 360 degrees in a circle.
1 degree = 60 arc-minutes
1 arc-minute = 60 arc-seconds
1 degree ≈ 0.0175 radians (rad)

DISTANCES
1 kilometer (km) ≈ 0.621 mile (mi)
1 astronomical unit (AU) ≈ 149,598,000 km ≈ 93 million mi
1 AU ≈ 8.3 light-minutes
1 light-year (ly) ≈ 63,241 AU ≈ 9.46 trillion km
1 ly ≈ 5.87 trillion mi
1 parsec = 206,265 AU ≈ 3.26 ly

The top two figures explain the two types of comparisons found in the book: one from Chapter 1 (top), where apparent, or angular, sizes are compared, and one from Chapter 6 (middle), where celestial objects are compared with other objects at the same scale. The bottom figure features a call-out fact, or "FYI," as well as a cross-eyed stereoscopic view.

Angular Size Comparison

Actual Size Comparison

Stereoscopic Image

Step Four

HERE WE SHOW all known objects in the solar system larger than 254 kilometers in diameter—objects too large to have been shown completely in the previous picture. At this scale, Earth has a diameter of 12.8 millimeters, or about half an inch.

Peeking in from the right edge is the sun. The sun's diameter on this scale is 1.39 meters. It is so large a million Earths could fit inside. The sun is a ball mostly of hydrogen gas. In its core, hydrogen burns into helium, producing the energy that will allow it to shine for about 10 billion years. You can see sunspots on its surface; these are of an order the size of Earth.

The four rocky terrestrial planets, Mercury, Venus, Earth, and Mars, are shown. So is our moon. Mars's two moons are too small to be shown—we see them in the previous picture. A swath of rocky asteroids appear, including the largest—Ceres. The four gas giant planets—Jupiter, Saturn, Uranus, and Neptune—are displayed with their large moons. Jupiter's moon Ganymede and Saturn's moon Titan are both larger than Mercury. Saturn's dramatic rings are made of myriad tiny icy moonlets (from microscopic to perhaps the size of houses) orbiting it. The rings of Uranus are also visible. Jupiter and Neptune have rings as well—too faint to be seen here. Jupiter's red spot, a centuries-old storm, is larger than Earth is.

Known icy Kuiper belt objects are shown, the most famous being Pluto and its moon Charon. This picture explains why little Pluto lost its planetary status in 2006, being demoted to dwarf planet. It was the discovery of many other icy bodies like it in the same region, including a larger one, Eris.

We also show the nearest star, after the sun—the red dwarf star Proxima Centauri, as well as the nearby white dwarf star Sirius B. Sirius B was the first white dwarf star studied. It orbits Sirius, the brightest star in the sky. Since Sirius B is very small it is relatively dim. When the sun has used up its hydrogen fuel, after a red giant phase, it will eventually shrink to become a white dwarf. A forming white dwarf more massive than 1.4 solar masses will collapse to form a neutron star or black hole.

We have gathered these objects together in one place to compare them in size. At this scale the moon is about 38 centimeters from Earth, the sun is about 0.15 kilome-

FYI At the scale of 1:1 billion, Earth is the size of a marble and Jupiter is about the size of a large grapefruit.

ters away. Proxima Centauri is about 40,000 kilometers away, and Sirius B is more than 80,000 kilometers away.

The dramatic 3-D image of Saturn and its rings (opposite) is made from two pictures of Saturn by the Hubble Telescope at different times. Particles orbiting at the inner edge of the Cassini division (the dark, narrow gap in the rings) are in a two-to-one resonance with Saturn's moon Mimas, circling Saturn twice every time Mimas circles once. (View cross-eyed for 3-D. See pages 98-99 for instructions on cross-eyed viewing.)

Periodic perturbations by Mimas cause particles orbiting just outside this radius to spiral outward, partially clearing the Cassini division of particles and making it appear dark. In this picture we can see clearly the A ring (outside the Cassini division), the white B ring (inside the Cassini division), and inside that, the translucent C, or Crepe, ring which can be seen against Saturn's disk.

Two views of Saturn taken by the Hubble Space Telescope at different times. They give slightly different viewing angles and can be combined to make a dramatic 3-D image.

The standard theory of Saturn's rings has been that they were formed when an errant moon wandered too close to Saturn (inside what is called the Roche limit) and was ripped apart by tidal forces. But Saturn has no wandering moons. All its regular moons out to Titan are composed primarily of water ice just like the rings and are in nice circular orbits.

We (Gott and Vanderbei) have argued, along with our colleague Edward Belbruno, that Saturn's rings as seen today are the remnant of an originally much larger ring system. Outside the Roche limit, icy moons were able to form from the icy ring particles, but inside the Roche limit, tidal forces prevented the formation of large moons,

leaving the rings we see today. In this model, the rings would be old. Recently, the Cassini spacecraft has found new evidence suggesting that the rings are indeed old. The rings are relatively clean of contamination by meteoric dust, which suggested a more recent origin. The Cassini spacecraft, however, found that the rings to be substantial enough to dilute the dust, with particle collisions continually breaking apart particles, uncovering new, fresh ice surfaces relatively clean of dust. Thus, the rings can be old.

APPAREL

CHAPTER 1

NT SIZES

THE SHAPE OF THE SKY

APPARENT SIZES ·· 22

GO OUT TO A DARK AREA away from city lights some night and take a look. What is the shape of the sky? The sky looks like a giant dome, and Earth appears as a flat disk bounded by a circle—the horizon. The sky arching overhead forms a hemisphere. You are at the center of this hemisphere.

Walk ten miles, and the dome of the sky follows you. At the end of the trip, it still seems centered on you. "No matter where you go, there you are," as the science-fictional hero Buckaroo Banzai would say. But you're not really seeing the whole sky—part of it is blocked by Earth.

Put on a space suit. Now imagine that Earth suddenly vanishes. You are alone, floating in space. Look down, and you will see the rest of the sky—the other hemisphere previously blocked out by Earth. The sun can now be seen—it was just below the horizon and blocked by

Earth before. You are now surrounded by a complete sphere of stars. The sky is shaped like a sphere, and thus it is often referred to as the celestial sphere. Think of the celestial sphere as a big balloon. You are standing in the middle. Since you are in the middle of the balloon, no matter which way you look, you see its inside surface. At first, the ancients actually thought the stars were attached to the inside surface of a rotating crystalline sphere, but today we recognize that the celestial sphere just represents all the different directions you can look out in three-dimensional space. Since we can't tell by looking how far away the objects in the sky are, they all appear to be equidistant, and this illusion makes it seem that we're sitting inside a gigantic sphere of stars.

Only six individuals in human history have had a chance to take in this view of the whole sky at once. These were the Apollo astronauts who ventured outside their capsule on their way home from the moon, and even they didn't get a really good look, owing to the glare from their brightly illuminated spacecraft.

It has been a long time since the ancient Greek astronomer Anaximander (610–547 B.C.) first understood that the sky really had the appearance of a complete sphere, but even today only distant spacecraft such as the Wilkinson Microwave Anisotropy Probe (WMAP)—a NASA Explorer mission now at a location four times as far away as the moon—have been able to observe it in its full glory.

The globe at left projects the night sky onto the surface of a sphere, what we refer to as the celestial sphere. We on Earth are inside that sphere looking out at the stars. Because Earth is rotating, the celestial sphere appears to rotate overhead, leaving star trails in a long-exposure photograph (opposite).

A Naked-eye View

Our galaxy, the Milky Way, is shaped like a disk, or dinner plate, and our solar system is situated in the plate itself about halfway out to the rim. If we look out in the plane of the plate, we see a faint glow circling the sky along the galactic equator. This glow emanates from the light of a hundred billion stars that make up the disk of our galaxy but that are too faint for us to make out individually. The galactic equator in the sky is tipped at a large angle with respect to Earth's Equator. On the page opposite, we show a photograph of a naked-eye view of the Milky Way, taken from a dark site. At the center of the bulge in the picture is the galactic center, which lies at the center of the plate.

On the following two pages, we show a naked-eye view of the whole sky as seen away from city lights. It covers the entire celestial sphere. From left to right, this image represents a 360-degree panorama around the entire sky. The faint band of light from the Milky Way, which circles the galactic equator on the celestial sphere, is seen as a straight horizontal band. The galactic center lies at the center of the picture, and dark bands of interstellar dust can be seen obscuring parts of the galactic disk. The two smudges below the plane of the galaxy are the Large and Small Magellanic Clouds, satellite galaxies of the Milky Way.

If we look out in directions above and below the plate, we see nearby stars, those right in our own neighborhood within the plate—a sprinkling of about 8,000 stars visible to the naked eye and covering the whole celestial sphere. Thus the familiar constellations are all made up of relatively nearby stars located in our little neighborhood of the galaxy; they are bright because they are close.

The entire area of the sky is divided up into 88 constellations, each with borders like those between countries on Earth. If a new object is found, it can be named after the constellation in which it resides. Thus, the black hole

Cygnus X-1 is named after the constellation Cygnus, in which it is found. The famous Andromeda galaxy is in the constellation of Andromeda.

A number of constellations may already be familiar to you. Orion, the Hunter, is one you may have seen on a cold winter's night. The Big Dipper is part of the constellation of Ursa Major, the Great Bear. We have outlined these two on the map of the sky (pages 26-27). We have also labeled some famous bright stars such as Vega and Sirius. The sun, the moon, and the planets (which continually move around the celestial sphere) are not shown, but they can always be found in the constellations of the zodiac.

FYI Almost all of the objects you can see with your naked eye in the night sky are within our Milky Way galaxy.

The 12 constellations of the zodiac form a circle around the sky through which the sun happens to journey during the course of the year as we move around it. You probably know your sign of the zodiac. The sun is supposed to be in that constellation on the date of your birth, but the zodiacal signs have not been updated in about 2,000 years, so your sign shows where the sun *would* have been on your calendar birth date a couple of thousand years ago but not today. There is, however, another particular constellation associated with you: At the moment you were born, straight up above you there had to be one constellation. For us, the authors, these constellations are Andromeda (Gott) and Hercules (Vanderbei)—our "hometowns" on the celestial sphere.

The band of light known as the Milky Way is at its widest when one looks at it in the direction of the galactic center.

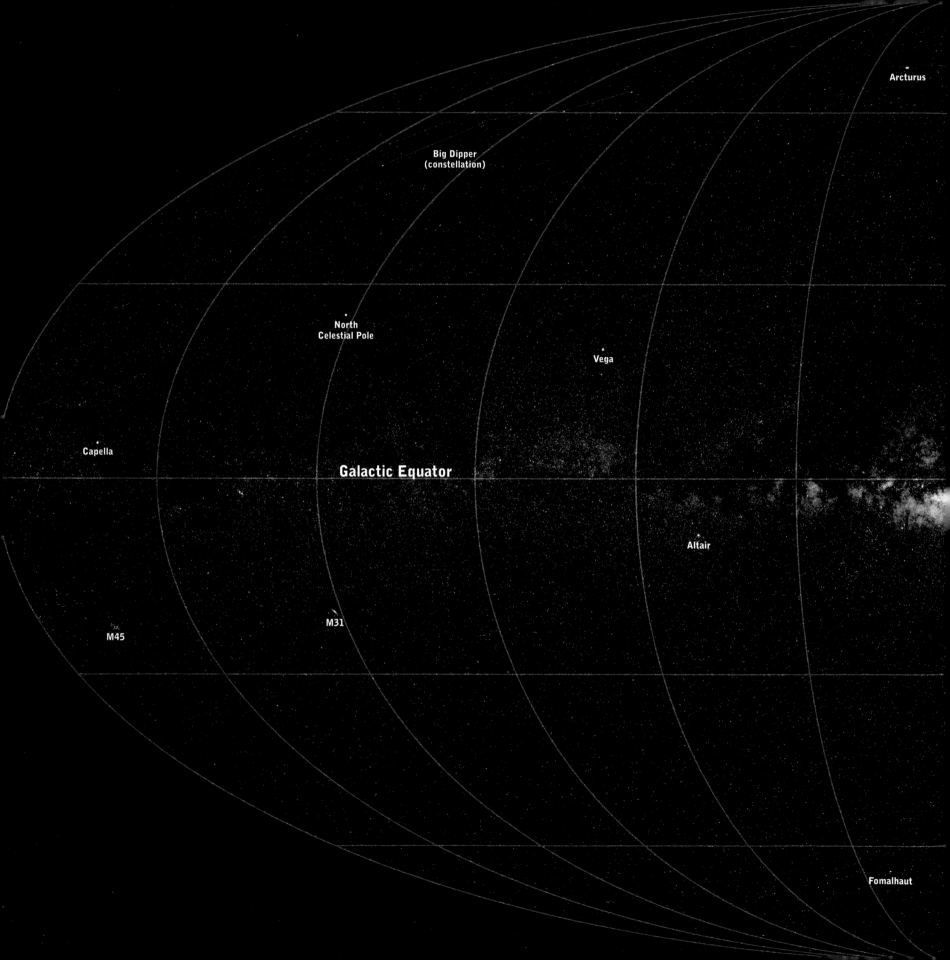

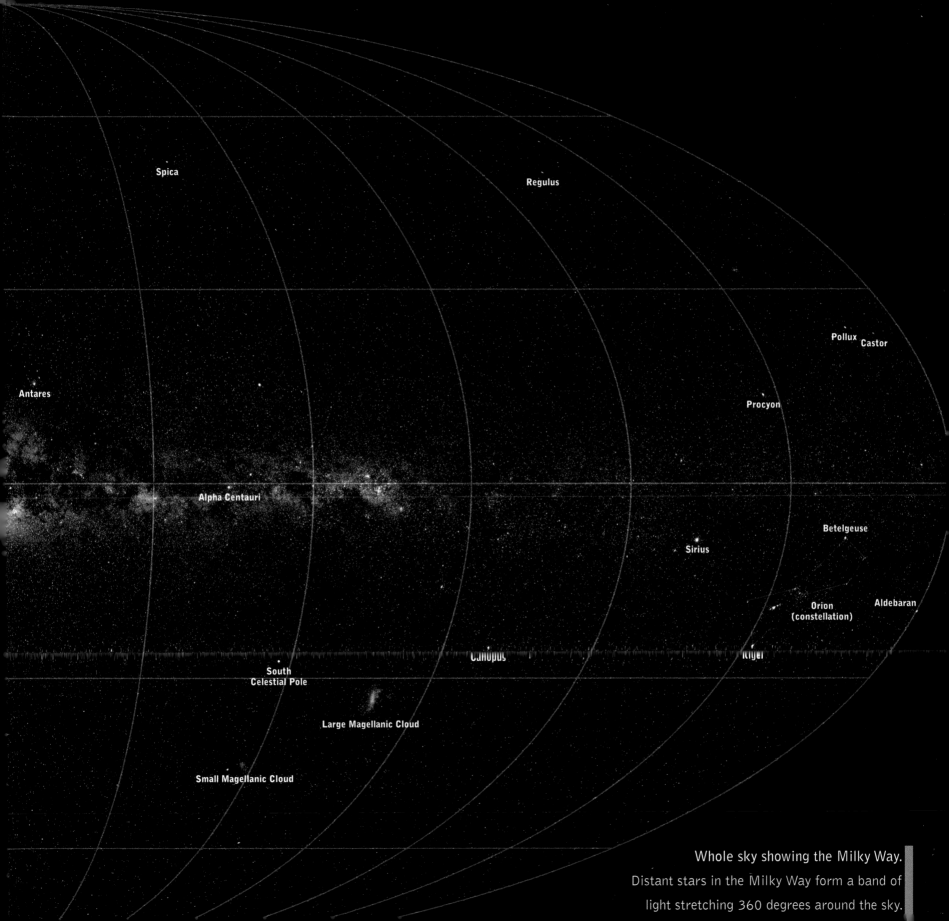

Spica

Regulus

Pollux Castor

Antares

Procyon

Alpha Centauri

Betelgeuse

Sirius

Orion
(constellation)

Aldebaran

Canopus

Rigel

South
Celestial Pole

Large Magellanic Cloud

Small Magellanic Cloud

Whole sky showing the Milky Way.
Distant stars in the Milky Way form a band of
light stretching 360 degrees around the sky.

Points of Reference

The sun and the moon are the same apparent size in the sky, each about half a degree across. Never is this more apparent than when the moon crosses exactly between Earth and the sun to cause a solar eclipse. The moon's orbit around Earth is not a perfect circle, so its distance from Earth changes a little during its orbit each month. When it is farther away, it looks smaller, and when it is closer, it looks bigger.

If the moon crosses in front of the sun while at its greatest distance from Earth, it causes an annular solar eclipse, as shown in the photograph opposite. Being a bit farther away than usual, the moon is slightly smaller in apparent size than the sun. Although it almost completely blots out the sun, it leaves a bright ring, or annulus, visible around the edge. This annular eclipse allows us to gauge the relative angular sizes of the moon and the sun—or how big they appear to an observer on Earth. If the moon passes in front of the sun when the moon is closer to Earth, its image is larger and can completely cover the bright surface of the sun; we refer to this as a total solar eclipse. During the few minutes when the moon is covering the sun, one can see the sun's faint outer atmosphere, its corona.

Because the moon blots out the sun when it passes in front of it, the ancient Greeks knew that the moon must be closer to us than the sun is. If the sun is about the same apparent size as the moon in the sky but is farther away, then it must actually be bigger than the moon.

Apparent Size Matters

Apparent, or angular, sizes are important. The angular size of the sun plays a crucial role in determining the temperature of Earth. If the sun were just twice its current angular diameter (say, by being only half as far away), it would take up four times as much area in the sky, and it would deliver four times as much sunlight to Earth. That would be like having four suns up in the sky! Earth would become a lot hotter.

With Earth's current reflectivity and atmospheric greenhouse effects, if the sun were twice as large in angular diameter in the sky, the Earth's average temperature would increase by 126°C (227°F). That's above the boiling point of water: The oceans would boil. If the sun were half its current apparent size in the sky (by making it twice as far away), the temperature of the Earth would drop by 84°C (151°F), and the oceans would freeze. So it's no accident that the sun is the size it appears: If it were much larger or smaller in the sky, we wouldn't be here to appreciate it.

Along these same lines, since the angular sizes of the sun and the moon are the same, their tidal effects on Earth would be equal if their internal densities were the same. However, the moon, which is made up of rock, is about 2.4 times as dense as the sun, which is made up of gas. Thus, tides produced by the moon are about 2.4 times larger than those produced by the sun. Ocean tides follow the moon but are modulated in amplitude according to the sun's position relative to the moon.

Sir Isaac Newton (1642–1727) figured out that ocean tides are caused by the gravitational effect of the moon. Because tides produced by the moon are stronger than those produced by the sun, he deduced correctly that the moon is denser than the sun. Newton's theory of gravity tells us that if the moon were twice as large in apparent size—by being twice as close—the ocean tides would be eight times as large, swamping coastal areas. Angular sizes are important indeed.

An annular solar eclipse occurs when the moon, at a distant point of its orbit, passes in front of the sun and almost, but not quite, covers it.

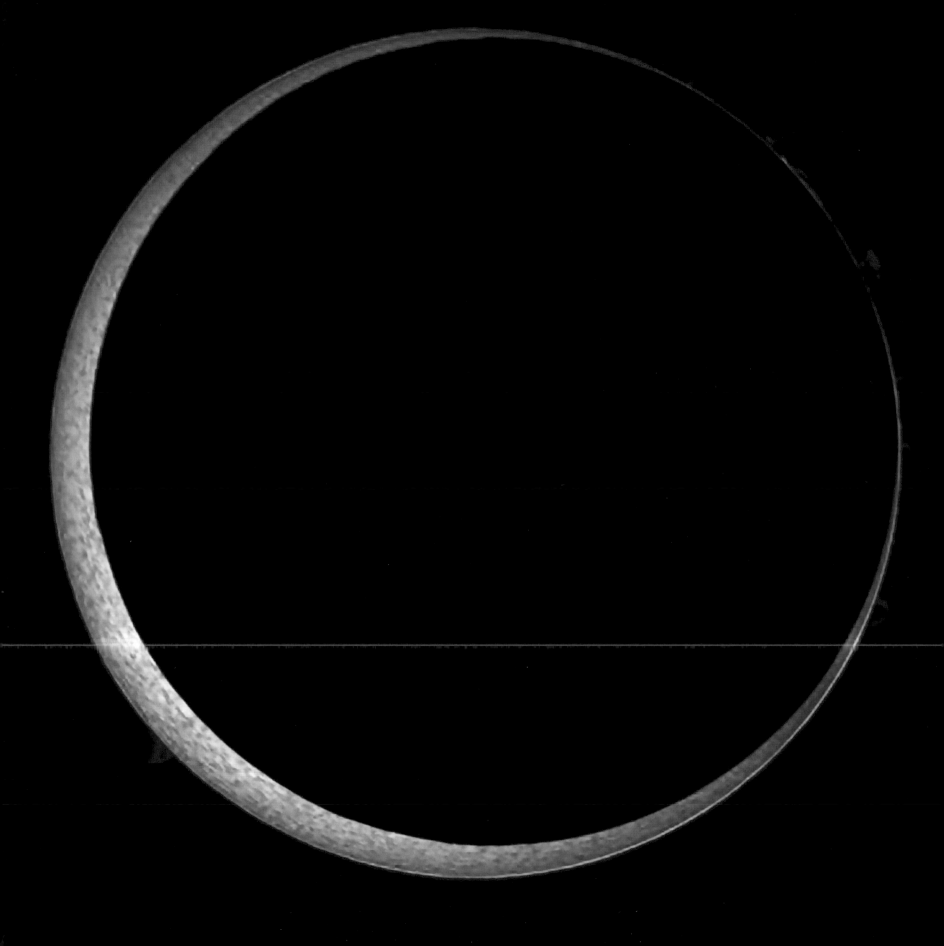

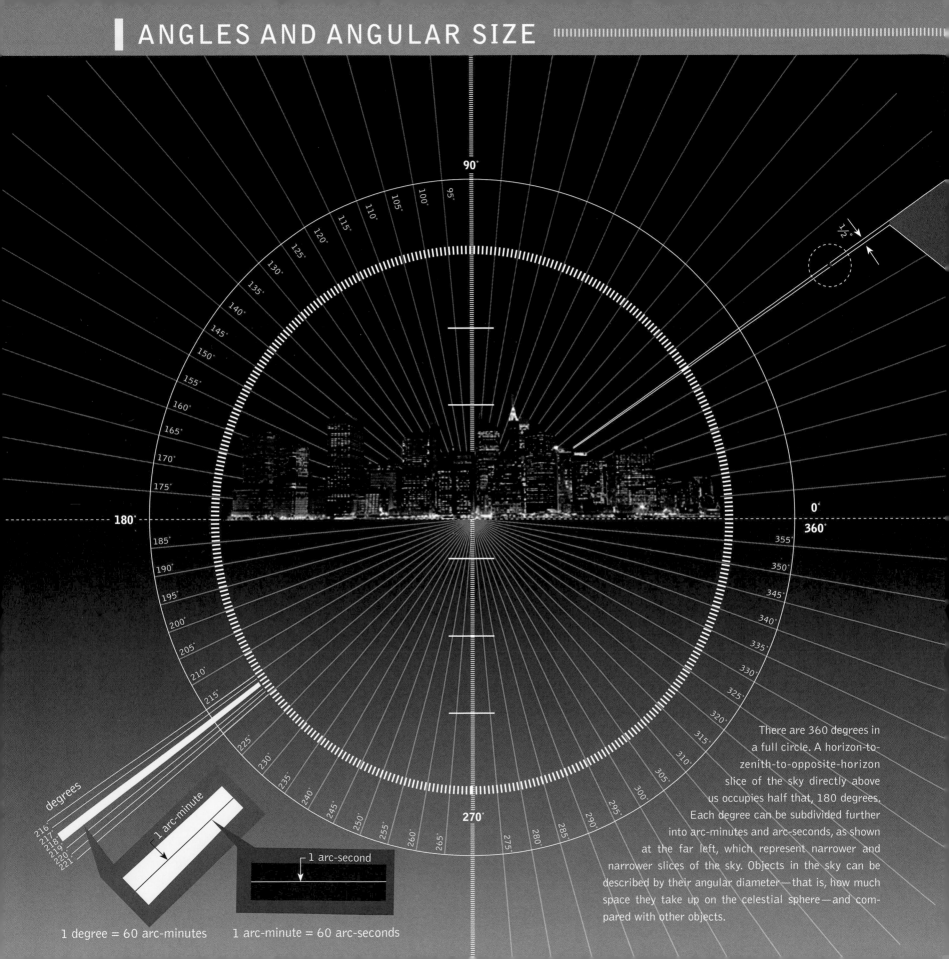

90°

95°
100°
105°
110°
115°
120°
125°
130°
135°
140°
145°
150°
155°
160°
165°
170°
175°

180°

185°
190°
195°
200°
205°
210°
215°
220°
221°
225°
230°
235°
240°
245°
250°
255°
260°
265°

270°

275°
280°
285°
290°
295°
300°
305°
310°
315°
320°
325°
330°
335°
340°
345°
350°
355°

0°
360°

½°

degrees

216°
217°
218°
219°
220°
221°

1 arc-minute

1 arc-second

1 degree = 60 arc-minutes   1 arc-minute = 60 arc-seconds

There are 360 degrees in a full circle. A horizon-to-zenith-to-opposite-horizon slice of the sky directly above us occupies half that, 180 degrees. Each degree can be subdivided further into arc-minutes and arc-seconds, as shown at the far left, which represent narrower and narrower slices of the sky. Objects in the sky can be described by their angular diameter—that is, how much space they take up on the celestial sphere—and compared with other objects.

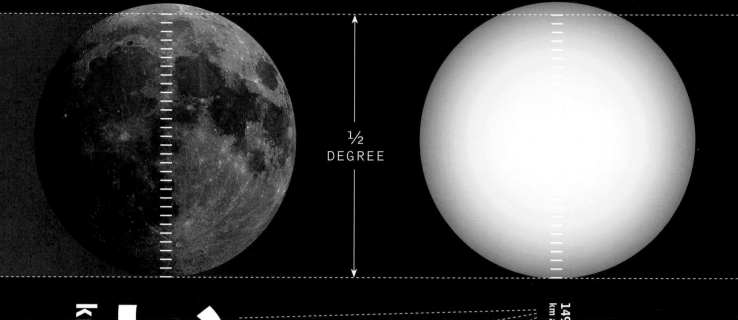

½ DEGREE

km away

384,400

149,600,000 km away

THE MOON AND THE SUN are the same angular size in the sky, a half degree, but the sun is 400 times farther away, which means that the sun's actual size is 400 times larger than the moon's.

Your hand held at arm's length provides convenient approximate angular measures. Your fist held at arm's length is about ten degrees wide, and your thumb is more than one degree wide. Your thumb held at arm's length will cover the moon. Try it—it's surprising.

The moon is much bigger than your thumb, of course, because it is much farther away. We typically learn the angular size of an astronomical object by taking a picture of it with a telescope of known focal length. If we can then find its distance, we can determine its actual size.

JUXTAPOSITIONS

IN THE NEXT FEW PAGES, we show a sequence of ever more magnified views of the sky with the moon arbitrarily placed as a template in each picture for comparison. In this way we can illustrate the angular sizes of various objects as seen from Earth. We begin by showing you how the moon would look if it ever wandered up near the Big Dipper, which it doesn't. The moon is just one-half degree in diameter. In contrast, the pointer stars of the Big Dipper (the two rightmost stars in this picture) are five degrees apart, or about ten lunar diameters.

FYI The fraction of the celestial sphere taken up by the Big Dipper is almost as large as the fraction of the surface of Earth taken up by the continental United States.

The apparent sizes of objects in the sky are measured by the angles they span in our field of view. A full circle spans 360 degrees. The horizon is a full circle on the celestial sphere, so a complete panorama around the horizon spans 360 degrees. You have to turn in a complete circle to view the entire horizon.

It was the ancient Babylonians who first divided the circle into 360 degrees. Since the sun has an angular size of half a degree (that's 1/720th of a full circle), it takes up just 1/720th of the horizon when it sets. Each day, the sun moves about one degree in the sky relative to the stars. As Earth circles the sun each year, the sun appears to circle the sphere of stars once a year. While a year is actually about 365.25 days, the Babylonians chose to divide the circle into 360 degrees because it is conveniently divisible by many numbers. Their definition of the degree stuck.

The moon may look large when seen setting against distant houses and trees along the horizon, but at arm's length, your thumb will still completely cover it. Half a degree is not very large. A typical digital camera in snapshot mode without zoom has an angular field of view of about 45 degrees. That means that it takes about eight pictures to stitch together a complete 360-degree panorama of the horizon. At a half degree wide, the moon takes up only 1/90th of the width of a typical digital camera field of view. That's why the moon may appear surprisingly small in your snapshots.

The moon is the astronomical object closest to Earth. Astronomers measure distances by how long it takes light, traveling at 300,000 kilometers (or 186,000 miles) per second, to cross those distances. The distance to the moon is 1.3 light-seconds.

When Apollo astronauts on the moon talked to President Richard M. Nixon, you could hear 2.6-second delays in the conversation. That is, it took the radio signals carrying Nixon's voice 1.3 seconds to travel to the moon and 1.3 seconds for the astronauts' replies to get back.

The stars in the Big Dipper are much, much farther away. Their distances range from 78 light-years to 124 light-years. That means that we are seeing these stars not as they appear today, but rather as they appeared 78 to 124 years ago.

The illustration opposite shows how the moon would look if it ever wandered up near the Big Dipper (which it doesn't). Viewed from a reading distance of about 17 inches, the angular scale is that of a normal naked-eye view from a dark site. The magnification is 1×, in other words, one times the magnification you see with the naked eye.

Big Dipper
78-124 light-years away

Near and Far

THE PICTURE OPPOSITE SHOWS the moon with the Rosette Nebula in the background. (The moon would never appear next to the Rosette Nebula in the sky; we are putting it here just to compare the apparent sizes of the two objects side by side.)

The Rosette Nebula is a gas cloud in our galaxy where stars are formed. It is 5,200 light-years away, so we are seeing it as it appeared 5,200 years ago. You can see the newly formed stars glowing in the center, illuminating the gas, causing it to glow.

FYI If you had much larger eyes, you could see more than a dozen faint objects in the night sky that appear bigger than the moon.

The Rosette Nebula is more than a degree wide, so it has a larger apparent size than the moon. So why haven't you seen it in the sky at night? The answer is that it is very faint. Your eye is tiny and simply does not have the light-gathering power to see such a faint object.

On the following two-page spread is a similar picture of the moon placed in front of the Andromeda galaxy (also called M31 for its position in the Messier catalog of nebulae). It is the nearest large galaxy we have found outside the Milky Way. The Andromeda galaxy is 2.5 degrees across—five times as wide as the moon. From a really dark site, with your naked eye or through a small telescope you can see the Andromeda galaxy as a faint smudge on the sky, but you are seeing only the innermost, brightest regions, and therefore it looks smaller than the moon. You miss the faint outer regions that show up in this long-exposure photograph.

These two comparisons illustrate how one of the great controversies in astronomy in the 20th century arose. Gas clouds like the Rosette Nebula residing in our own galaxy resembled, in angular size and appearance, spiral galaxies like Andromeda. Harvard astronomer Harlow Shapley thought all these were just gas clouds within our own galaxy. Only when Edwin Hubble discovered faint Cepheid variable stars (intrinsically very luminous, pulsating stars) in the Andromeda galaxy did he realize that, for them to appear that faint, Andromeda had to be very much farther away and, therefore, very much bigger than any gas cloud in our galaxy. Hubble proved that Andromeda and the other spiral nebulae were actually entire galaxies like our own. The Andromeda galaxy is 2.5 million light-years away, about 500 times farther away than the Rosette Nebula.

The next spread in the sequence shows an unlikely event, Mars passing near Saturn with the full moon hovering nearby. Mars and Saturn are shown as they would appear at their points of closest approach to Earth, 3.1 light-minutes and 71.2 light-minutes, respectively.

In the background of this spread is the kind of detail the Hubble Space Telescope can see at the same magnification. Ten thousand faint galaxies were counted in this Hubble Ultra Deep Field survey. Considering the small fraction of sky that the survey covered, it is estimated that more than 140 billion such galaxies are within the range of this telescope if we could survey the entire sky.

The magnification in this picture is 18×, similar to what would be obtained with a high-power pair of binoculars. You could see the stars in the Rosette Nebula, but the gas would still be too faint to see with binoculars.

Rosette
5,200 light-years away

Moon
1.3 light-seconds away

22 arc-minutes

APPARENT SIZES : 36

Moon
1.3 light-seconds away

17 arc-minutes

Andromeda Galaxy
2.5 million light-years away

APPARENT SIZES | 38

Mars
3.1 light-minutes away

Moon
1.3 light-seconds away

15.5 arc-seconds

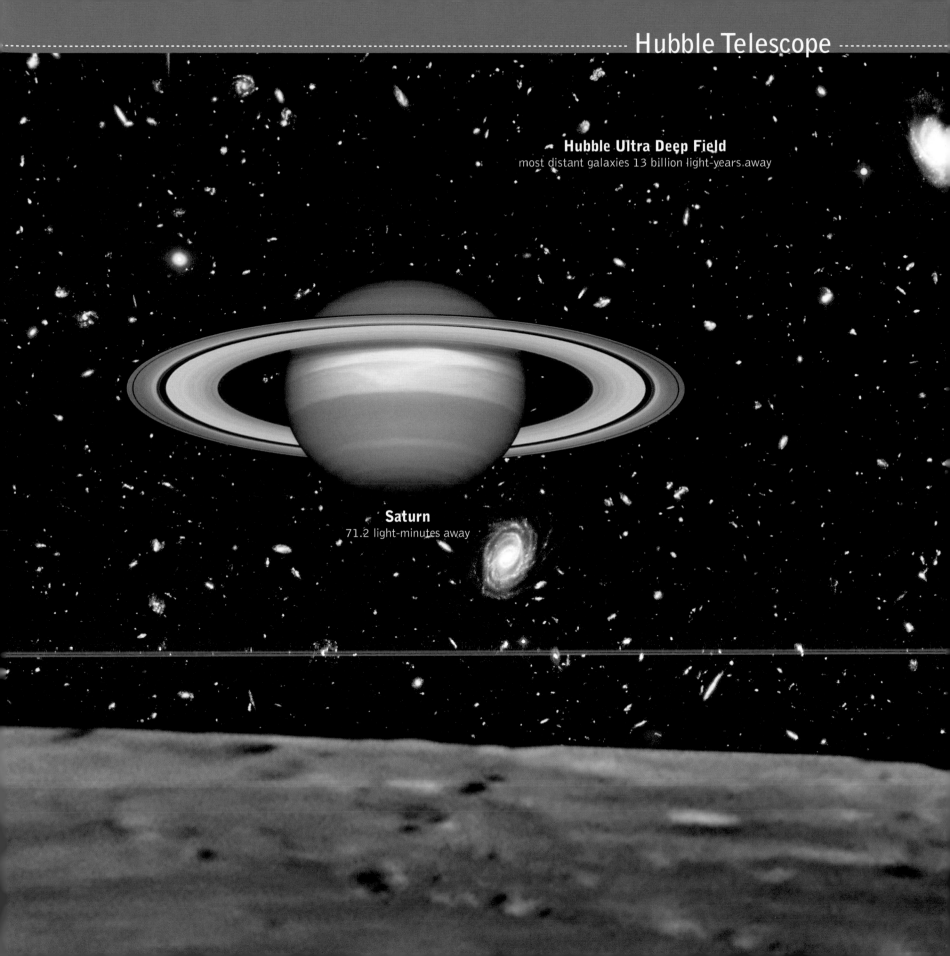

Hubble Ultra Deep Field
most distant galaxies 13 billion light-years.away

Saturn
71.2 light-minutes away

Starry Night

THE ILLUSTRATION OPPOSITE shows the angular sizes of stars as seen from Earth. The biggest star in angular size is Betelgeuse. This picture was taken in ultraviolet wavelengths by the Hubble Space Telescope. The star's angular size is about 1/30,000 that of the moon. The other stars' angular sizes are known from measurements made by interferometers (using pairs of telescopes with a wide baseline) or deduced from their colors and luminosities.

Alpha Centauri is just as big as the sun, but very far away, and therefore it doesn't take up much space in the sky. It takes up less space on the celestial sphere than a manhole cover takes up on the globe of Earth. Alpha Centauri has about the same surface brightness as the sun, but it appears tiny in the sky and therefore is very faint relative to the sun.

Alpha Centauri has the same color as the sun—white. The sun is not yellow as it usually appears in kids' drawings. It is yellow or red only when it is low in the sky and atmospheric scattering has knocked out the blue portion of the light. When it is high in the sky, the sun is too bright to look at. White light from the sun can be broken into all the colors of the spectrum: red, orange, yellow, green, blue, violet. To paraphrase Neil deGrasse Tyson: The sun is not yellow, it is white; if the sun were yellow, snow would be yellow. Except for Betelgeuse (whose ultraviolet image has had its color digitally enhanced), the stars are depicted in their true colors as they would appear to the eye. Our eyes are able to detect the colors of the brightest stars. For example, Sirius and Rigel are blue, Alpha Centauri is white, and Betelgeuse is red.

The 8,000 stars visible to the naked eye are scattered over the sky. Most of them are fainter than Alpha Centauri and appear with smaller angular sizes in the sky. Imagine looking at these 8,000 naked-eye stars scattered around the sky from the center of a transparent Earth. Imagine how little of the Earth's surface 8,000 randomly scattered manhole covers would cover. The naked-eye stars cover an even smaller fraction of the celestial sphere, so you can see how empty the sky is. We can see only out to a finite distance—13.7 billion light-years, to be exact, because light from farther away has not had time to reach us yet—and most lines of sight do not intersect a star. That's why the sky at night is dark.

FYI The stars we see at night are actually other suns—they appear faint only because they are so far away.

Since most lines of sight do not intersect a star, the line of sight pointing upward from your location at the time of your birth is not likely to intersect a star either. The chance that a naked-eye star was directly above your location at the moment of your birth is less than one in a trillion. You'd have to be really lucky for that to happen, meriting the designation of being "born under a lucky star."

At the magnification shown on the opposite page, our view of the moon now shrinks to include just a 100-yard patch of lunar soil. Here we can just make out the Apollo 11 descent module and its shadow, as seen from lunar orbit but scaled as they would appear from Earth.

This illustration juxtaposes the apparent sizes of some well-known stars together on a single page. For comparison, the Apollo 11 landing site is shown as it would be seen from Earth under the same high magnification.

APPARENT SIZES : 40

Arcturus
36.7 light-years away

Aldebaran
65 light-years away

Rigel
800
light-years away

Vega
25.3
light-years away

Sirius
8.6
light-years away

Alpha Centauri
4.4
light-years away

Betelgeuse
640 light-years away

100 yards

Apollo 11

Moon
1.3 light-seconds away

Antares
600 light-years away

0.025 arc-second

Aldrin's Footprint

FOR OBJECTS OF EVEN SMALLER angular size, we show an even more magnified view. Alpha Centauri can be seen peeking in at the side. Proxima Centauri, its companion, and the closest star to Earth, is also shown. It is a red dwarf star. The star HD 209458 is a star similar to our sun but considerably farther away than Alpha Centauri. It has a Jupiter–like planet that transits in front of it, which we show in the illustration opposite. Astronomers measured that planet's size by observing how much the luminosity of the star was dimmed as the planet passed in front of it.

FYI A telescope with a magnification of **14,000,000×** would be able to give you as good a view of a quarter from a distance of **4,000** miles as you get from holding it in your hand.

Finally, we can see Buzz Aldrin's footprint on the moon. It should still be there to observe. If we could get above the perturbing effects of Earth's atmosphere and if we had a telescope about 40 kilometers (25 miles) wide, we could see it this well.

We have now had a look at the entire sky. One way to put it all in perspective is to think of the celestial sphere as it might be mapped onto or overlaid on a globe of Earth. In this projection, the band of light we call the Milky Way would encircle it, tracing a circumference of some 40,000 kilometers (25,000 miles). The Big Dipper would be almost as large as the continental United States. The image of the Andromeda galaxy would be about the size of New Jersey. The images of the sun and moon would

each be 56 kilometers (35 miles) across, about the size of a large metropolitan city. Jupiter's image would be seven-eighths of a mile across. Alpha Centauri's image would be only 25 centimeters (10 inches) across, and Proxima Centauri's image would be 3.0 centimeters (1.2 inches) across.

Now that we understand the apparent sizes of objects in the universe as seen from Earth, we are on a better footing to discover their actual sizes, which we will come to in Chapters 5 and 6. But first, in the following chapter, we will show how and where one can find these objects in the night sky.

With 14,000,000-power magnification provided by a hypothetical supertelescope, we would be able to see dramatic views of nearby stars as well as the footprint Buzz Aldrin left behind on the moon.

14,000,000×

Buzz Aldrin's Footprint
1.3 light-seconds away

Proxima Centauri
4.2 light-years away

HD 209458
154 light-years away

Sirius
8.6 light-years away

0.00166 arc-seconds

Alpha Centauri
4.4 light-years away

MAPPING

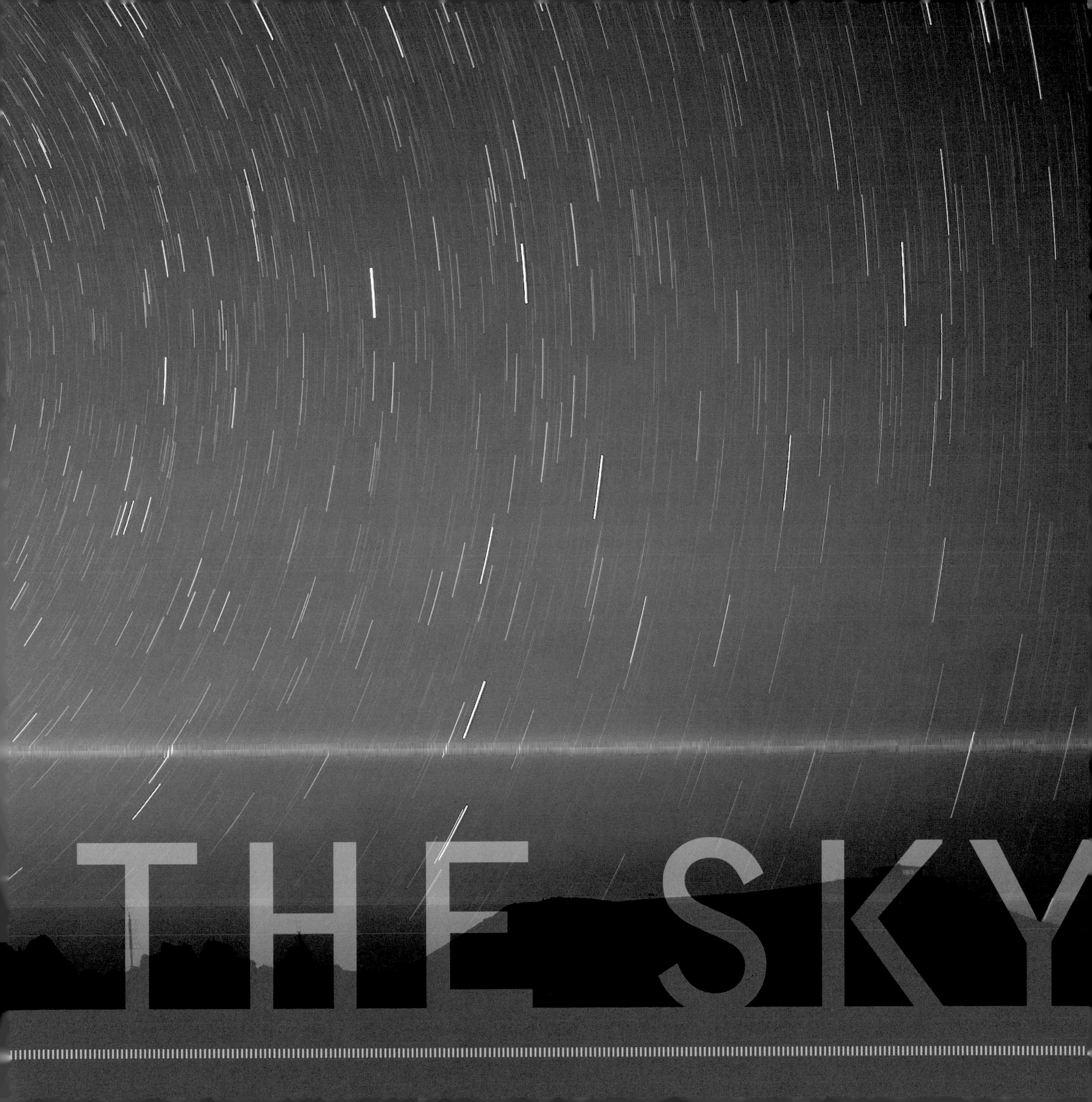
THE SKY

CHAPTER 2

MAKING MAPS

CARTOGRAPHERS MAPPING the Earth's surface (and the celestial sphere) were faced with the challenge of mapping a curved surface onto a plane. Such map projections aren't perfect, since some distortion is inevitable. However, they can capture important features. Perhaps the most famous map projection is the Mercator projection, developed by Gerardus Mercator (1512–1594) in 1569. This projection is conformal, which means that it correctly represents shapes in small areas. Lines of latitude are drawn as straight horizontal lines, and meridians of longitude are shown as straight vertical lines. Compass directions and angles are preserved over small areas, and shapes in local regions are correct. On Mercator's map, the shapes of both Iceland and South America are shown well. However, the scale varies significantly, depending on latitude: Areas in the far north or south are shown as too large relative to areas near the Equator. The classic example is Greenland. On Mercator's map, it looks about the same size as South America, whereas in reality Greenland occupies only about one-eighth the area of South America.

Other map projections preserve other properties. The Mollweide map projection is an equal-area map; that is, it accurately represents areas but does not preserve shapes. South America is a little too tall and skinny and Australia is also squashed.

Different map projections are good for different purposes. Astronomers mapping the whole sky have often used the Mollweide projection, as we did on the previous map of the sky (on pages 26-27), oriented so that the galactic equator is on the equator of the map. The WMAP team used it to show the cosmic microwave background (see pages 108-109). The equal-area property of this map is particularly useful for showing galaxy clustering (see pages 104-105), because such a map needs to show accurately how the number of galaxies per unit area in the sky varies from place to place.

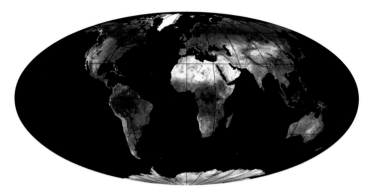

Mollweide map projection of Earth

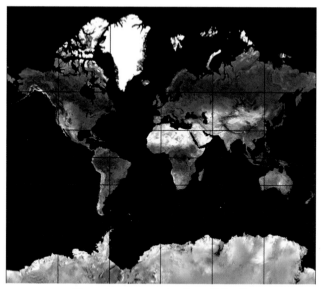

Mercator map projection of Earth

The shapes of the constellations in the Mercator map of the sky opposite are good, but ones in the far north (or south) like the Little Dipper (Ursa Minor) are too big relative to ones on the celestial equator like Orion.

Making a Celestial Cube

THE ONLY MAP PROJECTION of the celestial sphere on a flat piece of paper that can give a perfect view *as seen by your eye* is the *gnomonic projection*. The gnomonic projection is an old projection, produced in the following way. Place a lightbulb in the center of a sphere and project the sphere onto a plane (a piece of paper) just touching the sphere (below left). Only half of the sphere can be projected onto a plane at one time.

Let's first consider a gnomonic projection of Earth (below right). The continents and latitude and longitude lines have their shadows projected onto the plane.

The shortest distance between two points on a sphere is a *great circle* route. This is any circle on the sphere (like the Equator) whose center coincides with the center of the sphere. The gnomonic projection projects great circles on the sphere as straight lines on the plane. Notice that

the Equator and the meridians of longitude, which are all great circle routes, are straight lines on the gnomonic map of Earth. Charles Lindbergh used a gnomonic map of the North Atlantic when plotting his first solo flight from New York to Paris because he wanted to fly on the shortest possible route between the two cities.

For a great circle, the lightbulb at the center of the sphere lies in the same plane as the great circle itself. Because planes intersect planes in straight lines, when the shadow of the great circle cast by the light at its center falls on the plane of the map, it makes a straight line. A circle of latitude above or below the Equator is not a great circle because its center does not coincide with the center of the sphere, but falls above or below it. Latitude lines other than the Equator are curved on gnomonic maps.

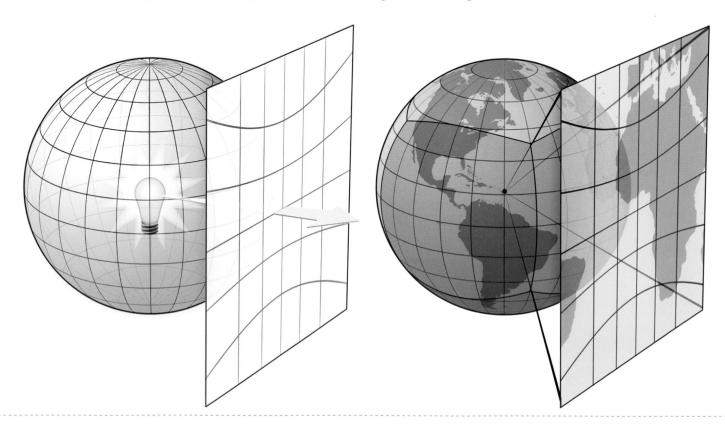

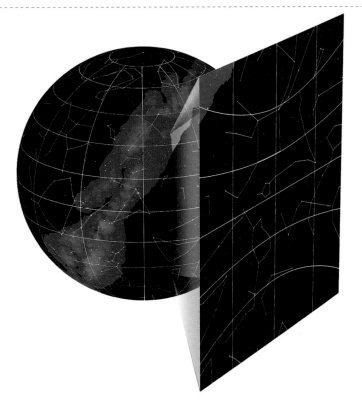

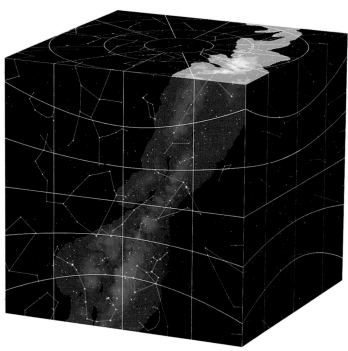

The celestial sphere can also be mapped onto a plane using the gnomonic projection (above left). This produces a flat star chart. The celestial equator is plotted as a straight line on this chart, as are all the meridians of celestial longitude (which astronomers call *right ascension*). Except for the celestial equator, circles of celestial latitude (which astronomers call *declination*) are shown as curved lines, like the shadows cast by the rims of a lamp shade on a wall.

If we place the celestial sphere inside a cube and put the light in the center, we can project the entire sphere onto the six faces of the cube. This produces a celestial cube that has a different star chart on each face (above right). When you look at a great circle on the sky, it looks straight—and it is straight on each face of the cube. But circles of celestial latitude (except for the celestial equator) are not great circles on the celestial sphere, do not appear straight in the sky, and are not plotted as straight lines on the faces of the cube.

This makes six star charts: The north circumpolar stars are shown in the top of the cube. The north celestial pole is in the center of the top. All the meridians of celestial longitude converge there and are straight lines on the top chart. Circles of celestial latitude appear as circles on the top chart. All stars north of 45° north celestial latitude are shown on this chart. These stars always remain above the northern horizon for an observer at 45° north latitude (for instance, northern U.S. or central Europe). The south circumpolar stars, stars south of 45° south celestial latitude, are all shown on the bottom chart. They are always below the horizon for an observer at 45° north latitude. The four sides of the cube form four star charts showing the stars that appear in each of the four seasons as seen by observers at mid-northern latitudes such as the United States: autumn stars, summer stars, spring stars, and winter stars. As Earth circles

The gnomonic projection (opposite left). Making a gnomonic map of Earth (opposite right). Making a gnomonic star chart (above left). Making a celestial cube (above right).

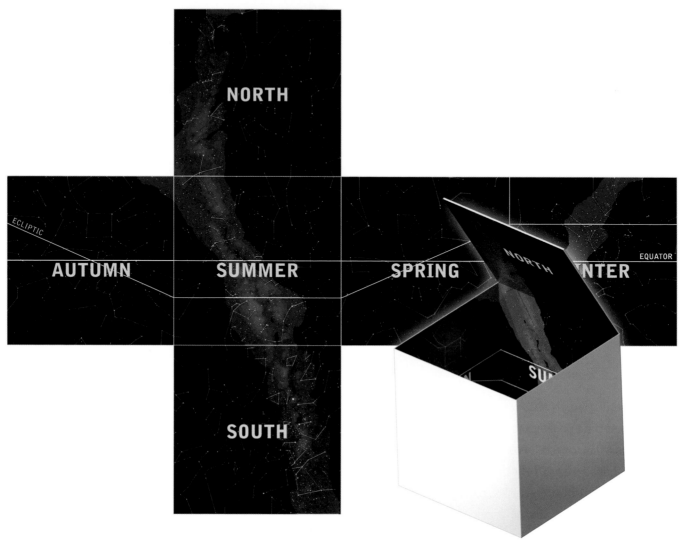

the sun over the course of the year, different groups of stars appear opposite it and are visible at midnight during each season.

Here a small version of the celestial cube star charts is pictured opened out into a cross-shaped configuration. The Milky Way is shown as a light band. The charts can also be folded up into a cube again with the charts on the inside, like the celestial sphere. Imagine standing in the center of that cube—the stars on the charts will appear to you just as they do in the sky. The white horizontal line is the celestial equator. The yellow line is the ecliptic, the path the sun takes through the sky during the year. Since

the ecliptic is a great circle, it is composed of a series of straight line segments on these gnomonic charts. Fold up the cube, and it makes a complete loop—like a rubber band around a box.

In the perspective view looking into the cube opposite, the North Stars are at the back of the cube and the south flap has been removed to allow a look in.

What's Above the Horizon?

HALF THE CELESTIAL CUBE is always above the horizon. If you are at mid-northern latitudes (approximately 45° north) and you go out at midnight on December 21, here is what you will see: Toward the north you will see the north circumpolar stars. The north celestial pole with Polaris, the North Star, will be about 45° above the northern horizon. The north circumpolar stars will extend from the northern horizon to the zenith (overhead). The winter stars will extend from the southern horizon to the zenith. The bottom of the winter star chart will sit on the southern horizon. The celestial cube will hang over you like a pup tent (below, and right). On the next page, in the top-left perspective drawing looking into the cube, as in the big diagram on the previous page, the part of the sky seen in winter at midnight is the black area enclosed in the trapezoid. Areas below the horizon

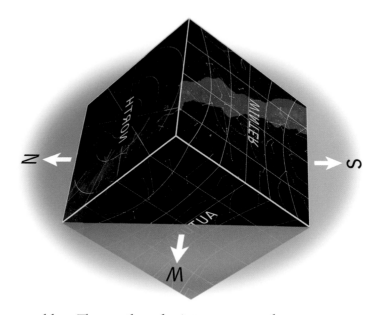

are blue. The north and winter stars can be seen, as can half of the autumn and spring stars. The four sides of the trapezoid represent the northern, western, southern, and eastern horizons, clockwise from bottom.

Similar effects occur in the other seasons as indicated opposite. The visible part of the sky rotates during the year. March 21 at midnight shows the north and spring stars, June 21 at midnight shows the north and summer stars. September 21 at midnight shows the north and autumn stars. The north circumpolar stars are always visible bordering the northern horizon.

A celestial cube oriented for winter observing from mid-northern latitudes (left) hangs like a pup tent over the observer, who is inside the tent on the ground at the center of it looking up. Half of the cube is "below the horizon" and so not visible. Black regions (opposite) represent parts of the celestial cube visible at different times of the year from 45° north latitude on Earth. The blue areas are below the horizon. In each case, as on the previous page, the perspective view is looking into the cube.

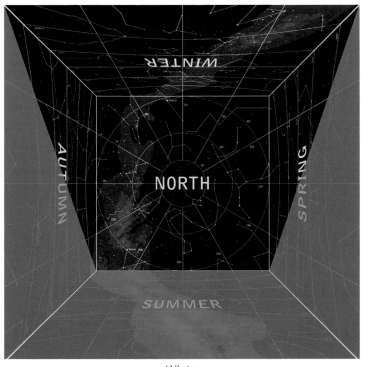

Winter

Spring

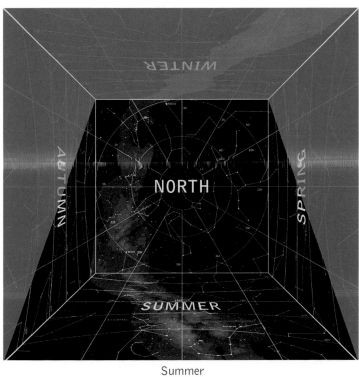

Summer

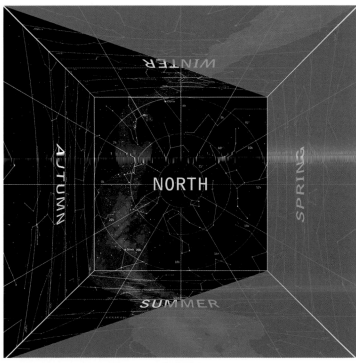

Autumn

ASTRONOMERS USE A SCALE called *apparent magnitude* to quantify brightness as seen by someone on Earth. For historical reasons, the brightest stars are called first magnitude, and the faintest stars visible to the unaided eye are called sixth magnitude: the lower the number, the brighter the star. In recent times, this scale was made more precise by declaring Vega to be magnitude 0 and declaring that sixth magnitude is 100 times fainter than first magnitude. Hence, each successive magnitude represents stars 2.512 times fainter than the one before (2.512 × 2.512 × 2.512 × 2.512 × 2.512 = 100). Objects brighter than Vega are given negative magnitudes. The sun is obviously the brightest thing in the sky—its magnitude is -26.7. The moon and planets are considerably fainter than the sun because they shine by reflected light. Stars are faint because they are very far away. Move a first-magnitude star 10 times as far away and it becomes 1/10th the angular size in the sky and covers 1/100th the area in the sky. It becomes 1/100th as bright. It becomes a sixth-magnitude star.

In this book, we make the important distinction between how big things *appear to be* versus how big they *are*. A similar distinction needs to be made here. Mercury and Sirius appear to be about equally bright. But Mercury is a planet within our solar system and Sirius is a star 8.6 light-years away—much farther away than Mercury. In absolute terms, Sirius is vastly brighter than Mercury. *Absolute magnitude* refers to the absolute, or true, brightness of an object. It is, by definition, the apparent magnitude of the object if it were seen from a standard distance of 32.6 light-years, or 10 parsecs, away. (A parsec is a unit of measure that is equal to 3.26 light-years; thus Sirius is 2.65 parsecs away.)

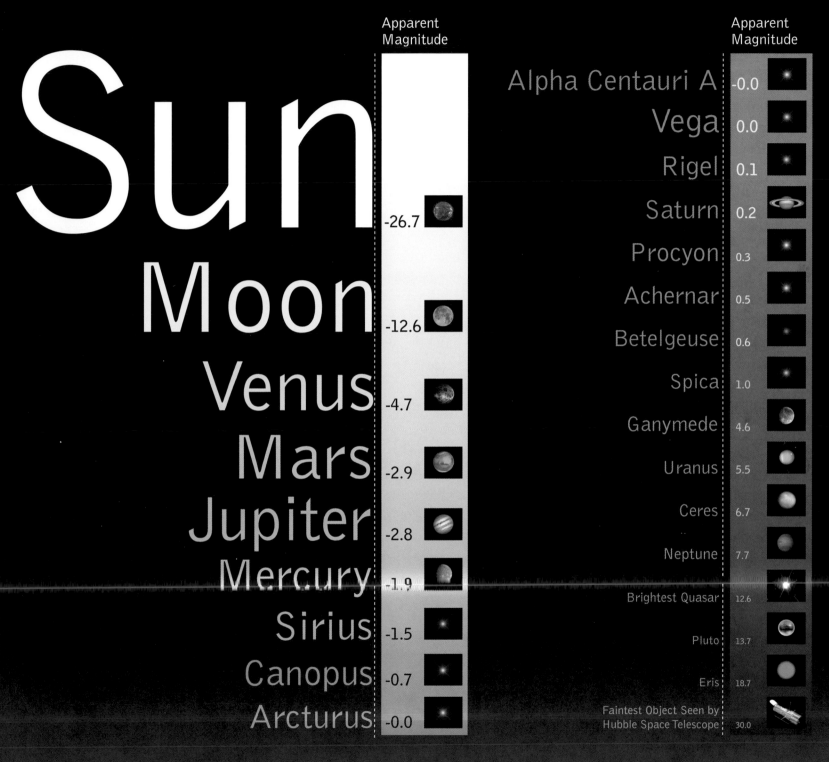

Apparent Magnitude (left)

Object	Apparent Magnitude
Sun	-26.7
Moon	-12.6
Venus	-4.7
Mars	-2.9
Jupiter	-2.8
Mercury	-1.9
Sirius	-1.5
Canopus	-0.7
Arcturus	-0.0

Apparent Magnitude (right)

Object	Apparent Magnitude
Alpha Centauri A	-0.0
Vega	0.0
Rigel	0.1
Saturn	0.2
Procyon	0.3
Achernar	0.5
Betelgeuse	0.6
Spica	1.0
Ganymede	4.6
Uranus	5.5
Ceres	6.7
Neptune	7.7
Brightest Quasar	12.6
Pluto	13.7
Eris	18.7
Faintest Object Seen by Hubble Space Telescope	30.0

TABLE OF BRIGHTNESS

The apparent magnitude of the brightest stars, planets, and select objects. For planets, the magnitude listed corresponds to its maximum brightness. For stars, absolute magnitudes—intrinsic brightnesses—are also given (right).

4.8	Sun	-6.7	Rigel
1.5	Sirius	2.7	Procyon
-5.5	Canopus	-2.8	Achernar
-0.4	Arcturus	-5.1	Betelgeuse
4.4	Alpha Centauri A	-3.5	Spica
0.6	Vega		

STAR CHARTS

THE FOLLOWING PAGES contain the six star charts forming the celestial cube. In each of these charts, constellations, famous stars, and other interesting objects are labeled. Each chart is ten inches by ten inches. If you put your eye exactly five inches in front of the center of one of these charts and hold the chart up, those stars' positions will appear to your eye exactly as they look in the sky. That is a property of the gnomonic projection we are using. Placing your eye five inches in front of the center of one of these charts places your eye exactly where it would be if it were in the center of the cube. Your line of sight to each star on the chart is just as it is in the sky. Since the sky rotates, you will have to rotate the chart until it lines up with what you see. The charts do not show the sun, moon, and planets because these move during the year, but they are always found close to the ecliptic—shown by a yellow line. The stars are so far away that their motions on the celestial sphere are imperceptible to the naked eye in a human lifetime.

FYI The angle between the north celestial pole and the northern horizon is equal to your latitude.

CELESTIAL COORDINATES: Just as places on the surface of Earth can be located by their latitude and longitude, objects on the celestial sphere can be located by their declination (celestial latitude) and right ascension (celestial longitude). Declination is measured in degrees north (+) or south (-) of the celestial equator. Right ascension is measured counterclockwise around the sky in hours, minutes, and seconds, where 24 hours equal 360 degrees. Lines of declination are labeled in degrees on the charts, and lines of right ascension are labeled in hours (for instance, 6h).

OBJECTS: For each chart, interesting objects are highlighted with small thumbnail portraits and brief descriptions. The coordinates for each object are given, with right ascension in hours *(h)* and minutes *(m)*, and the declination in degrees (°) and minutes of arc (')—for example, the Blinking Planetary (19h44.8m, +50°31').

NAMES: Objects are often named after their catalog number. The Whirlpool galaxy is also known as M51 because it is the 51st object in the Messier Catalog of Nebulae (which has 110 entries). The New General Catalogue (NGC) has 7,840 objects, expanded with two Index Catalogs (IC) having 5,386 objects. The Uppsala General Catalog of Galaxies (UGC) has 12,921 objects. The famous quasar 3C 273 comes from the Third Cambridge Catalog of Radio Sources. The Coma cluster comes from the Abell Catalog of Galaxy Clusters, so it is also called Abell 1656.

The Dumbbell Nebula (opposite), a dying star, is shown ejecting gas. Find it in the summer chart on page 71.

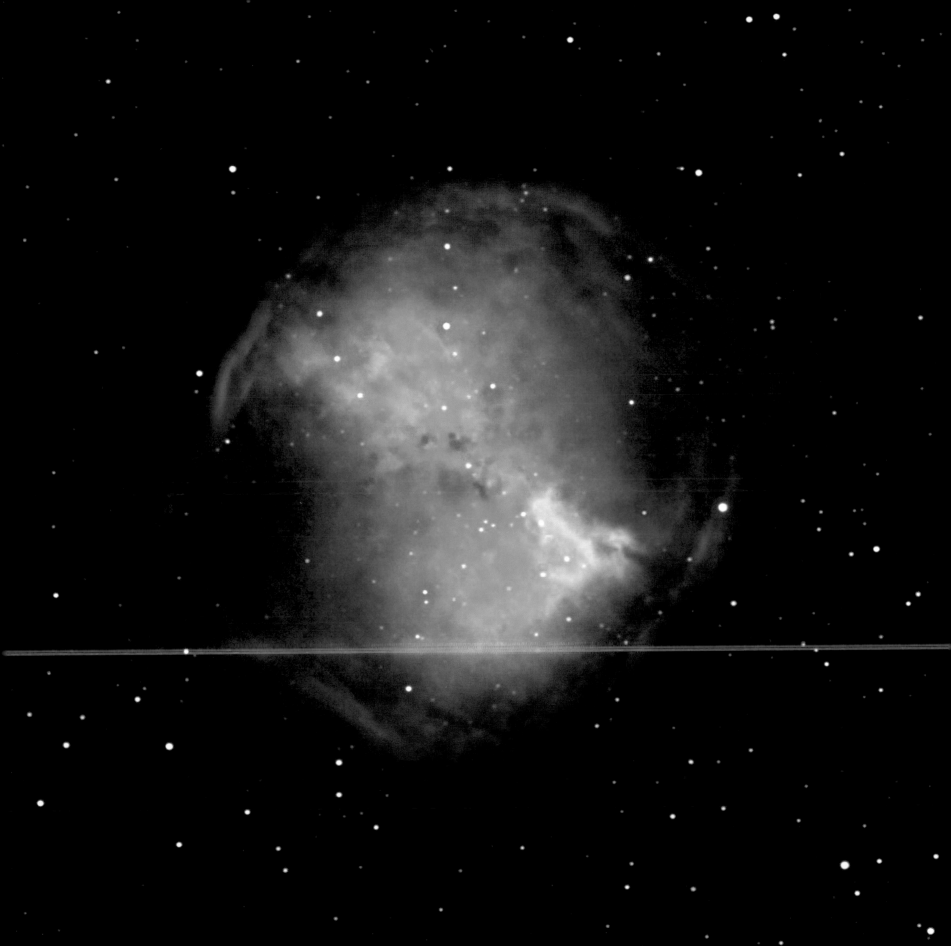

North Circumpolar

THIS CHART INCLUDES STARS that always remain above the northern horizon for an observer at 45° north latitude (for instance, the northern U.S.). At the center of the chart is the north celestial pole with the star POLARIS nearby. Follow the orange arrow from the pointer stars in the BIG DIPPER to find it.

BLINKING PLANETARY: A planetary nebula—a dying star ejecting gas.
NGC 6826
19ʜ44.8ᴍ, +50°31'

COCOON: Nebula includes regions of emission, reflection (gas reflecting starlight), and dust absorption.
IC 5146
21ʜ53.4ᴍ, +47°16'

NORTH AMERICAN: A huge emission nebula, the interesting "Mexico" part shown at left.
NGC 7000
20ʜ59.3ᴍ, +44°32'

BODE'S GALAXIES: Two bright galaxies: M81 is a spiral, whereas M82 is a starburst galaxy.
M81 AND M82
09ʜ55.6ᴍ, +69°04' See pp. 133, 227

DENEB: One of the most intrinsically luminous stars known—roughly 60,000 times brighter than our sun.
20ʜ41.4ᴍ, +45°17' See p. 209

OWL: A planetary nebula. The central star is magnitude 16.
11ʜ14.8ᴍ, +55°01'

BUBBLE: The central star in this nebula is blowing a giant bubble of material into space.
NGC 7635
23ʜ20.7ᴍ, +61°12'

DOUBLE CLUSTER: A beautiful pair of open clusters separated by only 28 minutes of arc.
NGC 869, NGC 884
02ʜ20ᴍ, +57°08'

PACMAN: An emission nebula.
NGC 281
00ʜ52.8ᴍ, +56°37'

BUTTERFLY: A glowing emission nebula. (Its gas glows like a neon sign.)
IC 1318
20ʜ26.3ᴍ, +40°07'

ELEPHANT TRUNK: An emission nebula.
VDB 142
21ʜ37.1ᴍ, +57°29'

PELICAN: An emission nebula.
IC 5070
20ʜ50.8ᴍ, +44°21'

CAPELLA: Eleventh brightest star in the night sky. Visually appearing as a single star, Capella is actually a binary star.
05ʜ16.7ᴍ, +46°00' See p. 26

HELIX: A rare "polar ring" galaxy.
NGC 2685
08ʜ55.6ᴍ, +58°44'

WHIRLPOOL: A spiral galaxy interacting with a satellite galaxy (NGC 5194), 23 million light-years away.
M51
13ʜ29.9ᴍ, +47°11' See pp. 134, 226

CAT'S EYE: A planetary nebula with complex structures such as knots and jets, and arcs.
NGC 6543
17ʜ58.6ᴍ, +66°38'

INTEGRAL SIGN: A spiral galaxy seen edge-on. The disk is oddly warped.
UGC 3697
07ʜ11.4ᴍ, +71°50'

CAVE: A faint emission nebula.
22ʜ56.8ᴍ, +62°37'

LITTLE DUMBBELL: A planetary nebula, one of the faintest in Messier's list.
M76
01ʜ42.4ᴍ, +51°34'

Map Key ● Galaxy ✦ Planetary Nebula ● Nebula (other than planetary) ✪ Globular Cluster ● Open Cluster

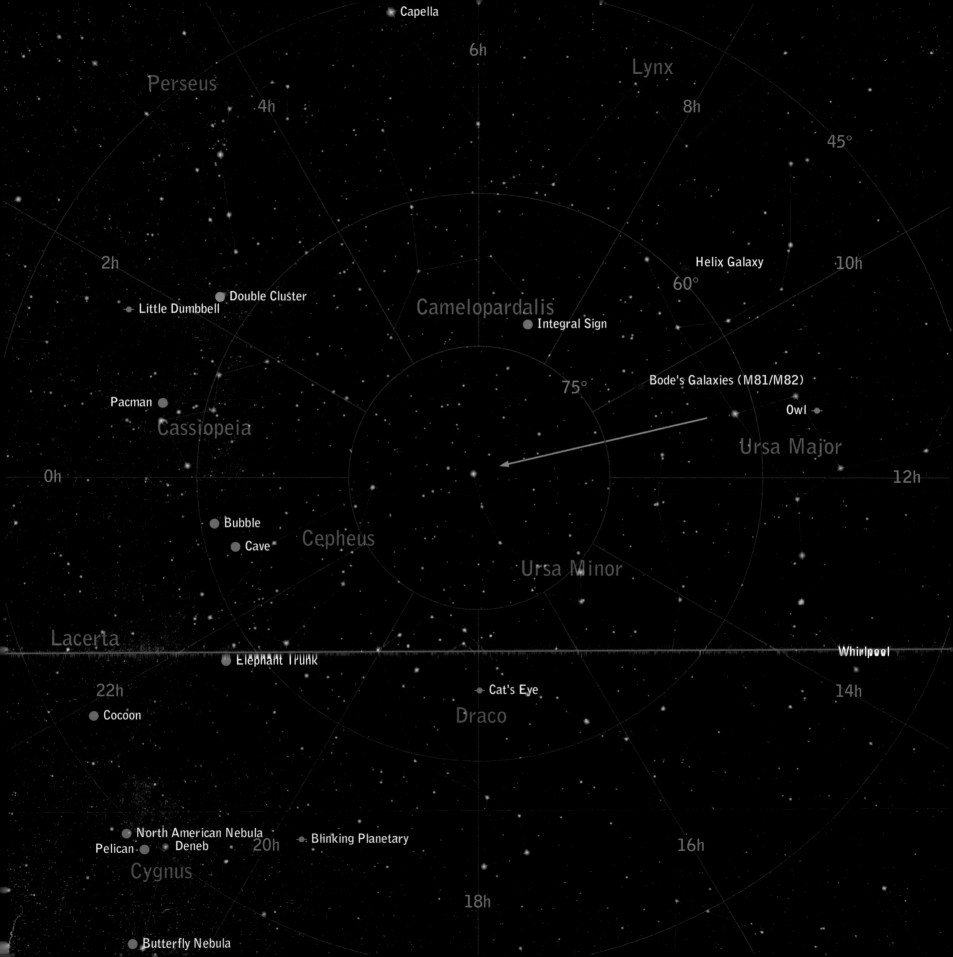

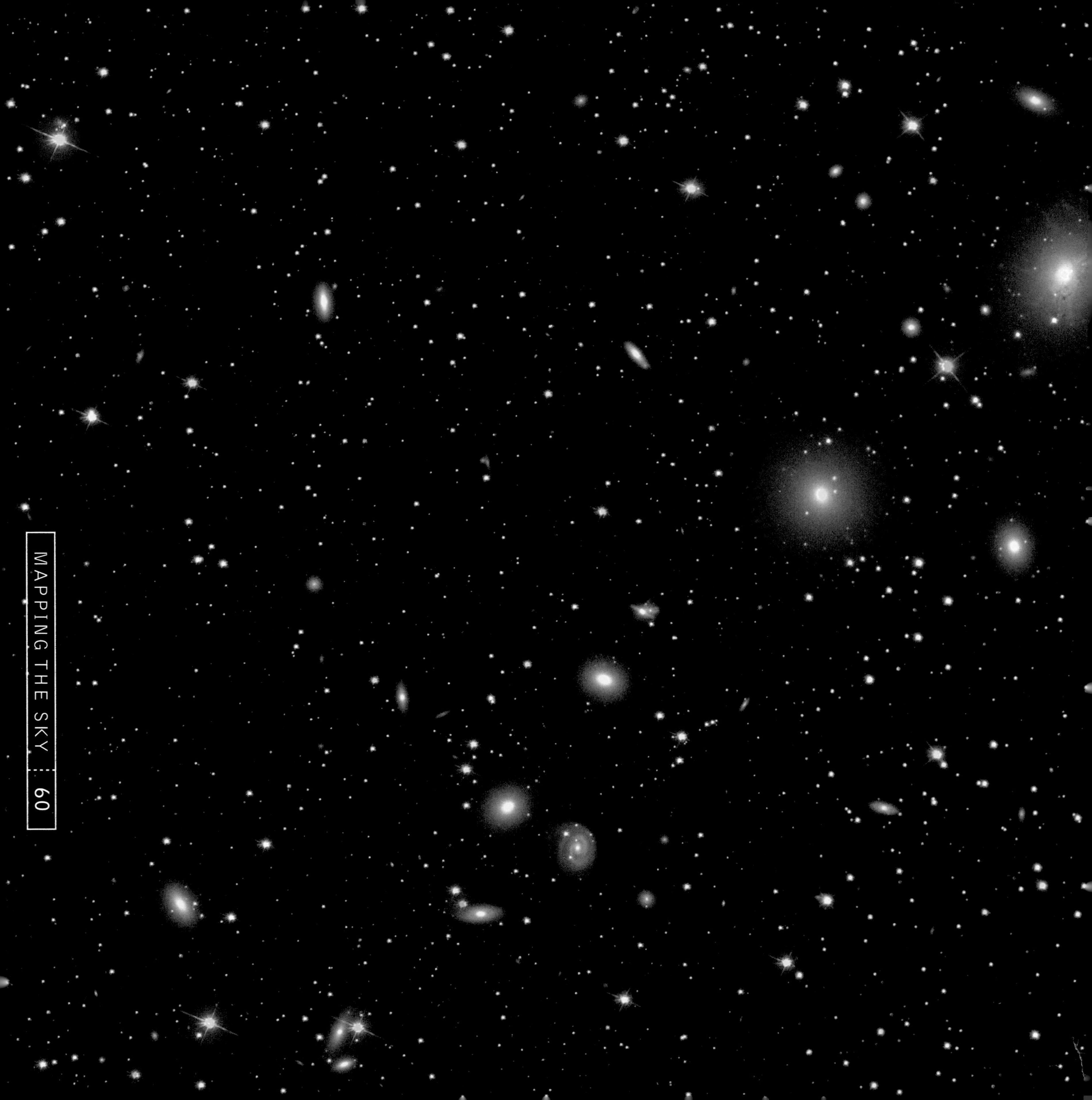

A spectacular image of the Perseus cluster of galaxies located in the North Circumpolar chart and captured here by the Sloan Digital Sky Survey.

Winter

AT MIDNIGHT ON DECEMBER 21 at mid-northern latitudes these stars are visible above the southern horizon. The ecliptic is the yellow line passing through the constellations of **TAURUS**, **GEMINI**, and **CANCER**. The sun passes through them during the summer but is on the opposite side of the sky during the winter, allowing viewing of these stars at night.

ALDEBARAN: The 14th brightest star in the night sky.
04H35.9M, +16°31' See pp. 27, 41, 209

HIND'S VARIABLE: A variable nebula. The central star varies in brightness; the nebula reflects its light.
NGC 1555
04H22.0M, +19°32'

POLLUX: The first "named" star known to have an extrasolar planet (a planet outside the solar system).
07H45.3M, +28°02' See pp. 27, 208

BEEHIVE: An open star cluster 577 light-years away.
M44, OR PRAESEPE
08H40.1M, +19°51' See p. 238

HORSEHEAD: A dark dust cloud nebula obscuring a more distant emission nebula, IC 434.
B33
05H41M, -02°28' See pp. 17, 216, 219

PROCYON: At 11.4 light-years, one of our closest neighbors. Magnitude 0.3. A binary star with a magnitude 10.7 companion.
07H39.3M, +05°13' See pp. 27, 55, 209

BETELGEUSE: A red supergiant, the ninth brightest star in the night sky.
05H55.2M, +07°24' See pp. 27, 40, 55, 132, 208, 210, 212, 214

HUBBLE'S VARIABLE: A variable nebula, reflecting the light of a variable star.
NGC 2261
06H39.2M, +08°45'

RIGEL: The seventh brightest star in the night sky at magnitude 0.1.
05H14.5M, -08°12' See pp. 27, 41, 55, 132, 209, 210

CALIFORNIA: An emission nebula. Very large, almost 2.5° long.
NGC 1499
04H03.3M, +36°25'

INTERGALACTIC WANDERER: A globular cluster about 300,000 light-years from Earth.
NGC 2419
07H38.1M, +38°53'

ROSETTE: A large, circular emission nebula with associated star cluster NGC 2244.
NGC 2237-2239
06H33.8M, +05°00' See pp. 35, 220, 240

CONE: An emission nebula embedded in the Christmas Tree Cluster.
NGC 2264
06H41.1M, +09°53'

MEDUSA: A very old and large planetary nebula.
07H29.0M, +13°15'

SIRIUS: Brightest star in the night sky (mag. -1.5). A binary star system.
THE DOG STAR
06H45.1M, -16°43' See pp. 27, 41, 43, 55, 96, 99, 202, 209, 210, 216, 219

CRAB: Remnant of the supernova from A.D. 1054, with a pulsar (neutron star) at its center.
M1
05H34.5M, +22°01' See pp. 132, 216, 219

ORION: A very bright diffuse nebula, the closest stellar nursery at 1,500 light-years away.
M42
05H35.3M, -05°23' See pp. 64-65, 132, 217

WITCH HEAD: Perhaps an ancient supernova remnant.
IC 2118
05H02M, -07°54'

ESKIMO: A bipolar double-shell planetary nebula.
NGC 2392
07H29.2M, +20°55' See pp. 216, 219

PLEIADES: Open star cluster with associated reflection nebulosity (gas reflecting the starlight).
M45, SEVEN SISTERS, OR SUBARU
03H47.4M, +24°07' See p. 132

Map Key Galaxy Planetary Nebula 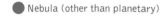 Nebula (other than planetary) Globular Cluster ⬤ Open Cluster

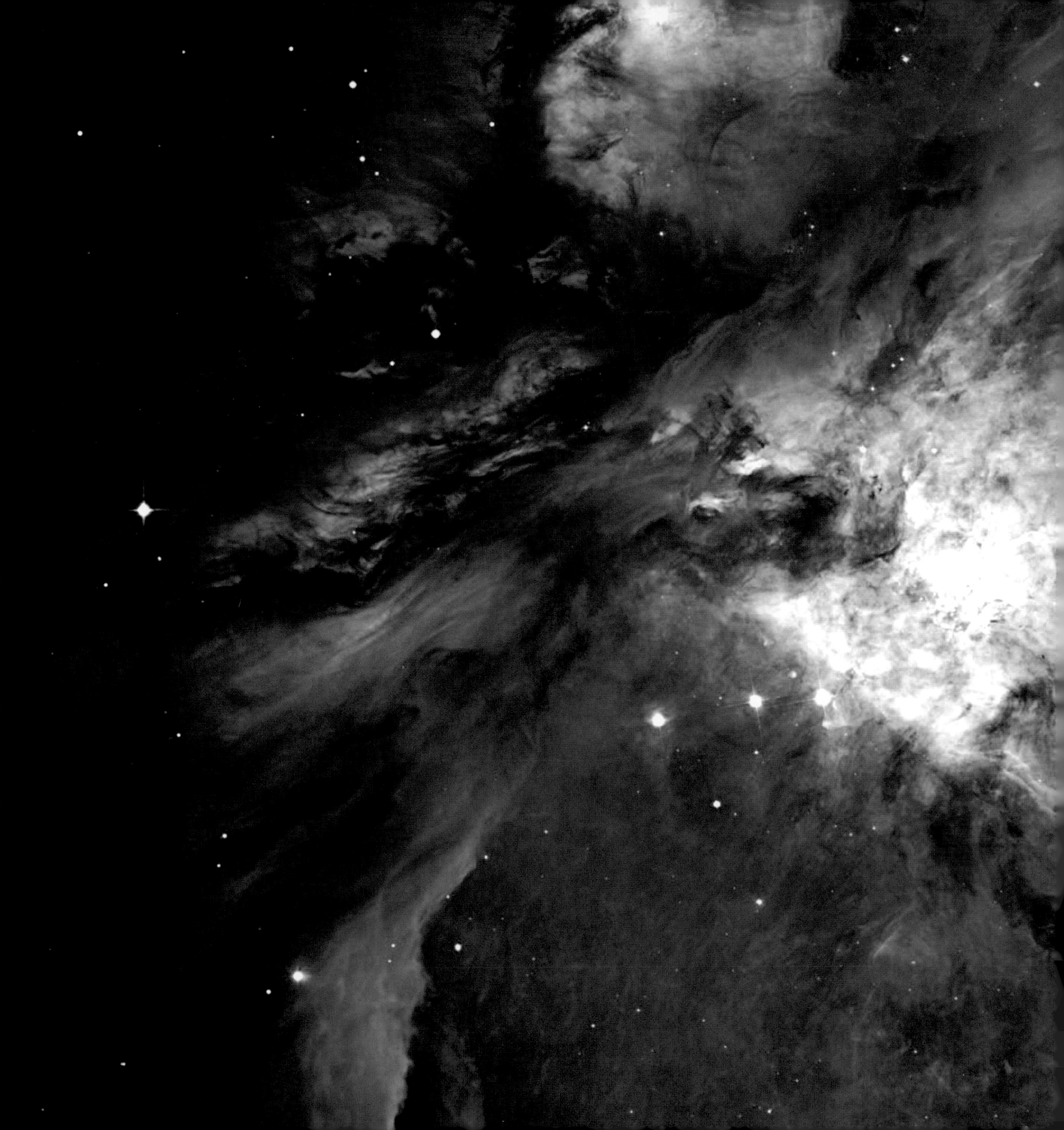

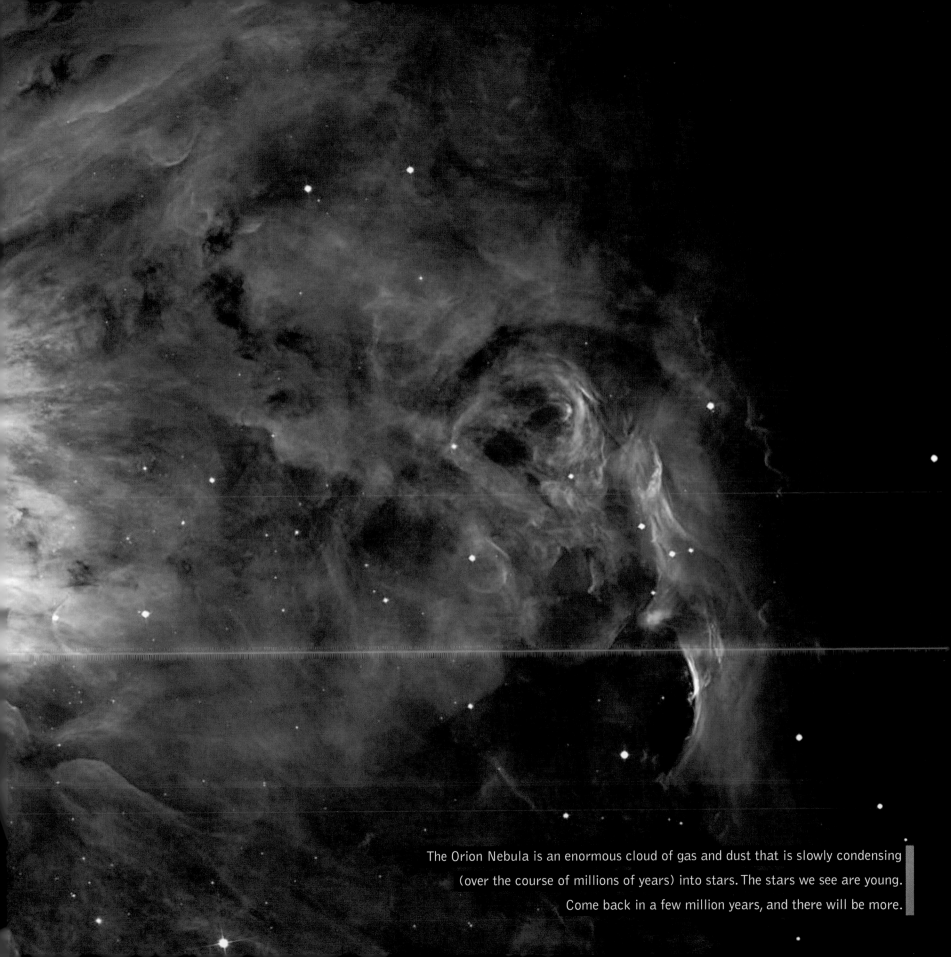

The Orion Nebula is an enormous cloud of gas and dust that is slowly condensing (over the course of millions of years) into stars. The stars we see are young. Come back in a few million years, and there will be more.

Spring

AT MIDNIGHT ON MARCH 21 from mid-northern latitudes you will see these stars above the southern horizon. The chart shows that the ecliptic crosses the **CELESTIAL EQUATOR** at an angle of 23.5°. That's because the axis of the Earth is tilted 23.5° relative to the line perpendicular to the plane of the Earth's orbit (the plane of the **ECLIPTIC**).

ANTENNAE: A pair of colliding galaxies. The collision is producing a pair of long tails of stars, gas, and dust.
NGC 4038 AND NGC 4039
12H01.9M, -18°52' See p. 133

LEO TRIO: A small group of three galaxies about 35 million light-years away.
M65, M66, NGC 3628
11H17M, +13°25'

SIAMESE TWINS: A colliding pair of spiral galaxies.
NGC 4567 AND NGC 4568
12H36.6M, +11°14'

ARCTURUS: The fourth brightest star in the night sky at visual magnitude 0.0. It is a red giant 36.7 light-years away.
14H15.7M, +19°11' See pp. 26, 41, 55, 209

M87: A giant elliptical galaxy in the Virgo cluster with an active nucleus. A "jet" of material is being ejected by a central black hole.
12H30.8M, +12°23' See pp. 212, 215

SOMBRERO: A bright spiral galaxy seen edge-on. Easily visible even with small telescopes.
M104
12H40M, -11°37' See pp. 68, 134, 226

BLACKEYE: A spiral galaxy. Inner region rotates in the opposite direction from that of the outer.
M64
12H56.7M, +21°41'

MARKARIAN'S CHAIN: A chain of more than a dozen galaxies.
M84, M86, ET AL.
12H27M, +13°10'

SPICA: At an average magnitude of 1.0, it dims slightly (by 0.1 magnitude) every four days as it and a companion star orbit each other.
13H25.2M, -11°10' See pp. 27, 55, 209

COCOON: An interacting pair of galaxies.
NGC 4490
12H30.6M, +41°38'

MICE: A colliding pair of spiral galaxies about 300 million light-years away.
12H46.2M, +30°44'

SPINDLE: A former spiral galaxy stripped of its gas. Seen edge-on.
NGC 3115
10H05.2M, -07°43'

COMA CLUSTER: A large cluster of galaxies.
ABELL 1656
12H59.8M, +27°59' See pp. 102, 134, 228, 231

NGC 4565: An edge-on spiral galaxy.
12H36.3M, +25°59'

SUNFLOWER: A spiral galaxy. Part of the M51 group.
M63
13H15.8M, +42°01'

COPELAND'S SEPTET: A group of seven galaxies.
NGC 3750
11H37.9M, +21°58'

QSO 3c 273: The first discovered and optically brightest quasar (mag. 12.6). This quasar has a jet.
12H29.1M, +02°03' See p. 134

GHOST OF JUPITER: A planetary nebula.
NGC 3242 OR CAT'S EYE
10H24.8M, -18°39'

REGULUS: Visual magnitude 1.4, young, massive, blue star 78 light-years away. It spins rapidly (once every 15.9 hours) and is therefore oblate.
10H08.4M, +11°58' See pp. 27, 91, 92, 209, 211

Map Key   ● Galaxy   ✦ Planetary Nebula   ● Nebula (other than planetary)   ✛ Globular Cluster   ● Open Cluster

NASA Hubble Space Telescope took this picture of the Sombrero galaxy (M104).

From our perspective this galaxy is seen almost edge-on.

It is 50,000 light-years in diameter and 28 million light-years away.

Summer

AT MIDNIGHT ON JUNE 21 from mid-northern latitudes these stars are visible above the southern horizon. The ECLIPTIC is the yellow line passing through SAGITTARIUS and SCORPIUS. THE MILKY WAY is prominent in the summer. THE RING NEBULA, the double star ALBIREO, and the DUMBBELL NEBULA are great objects to view with small telescopes.

ALBIREO: A beautiful binary star, one star orange, the other blue.
19H30.7M, +27°58'

ALTAIR: Visual magnitude 0.8. Period of rotation is 6.5 hours. Diameter at its equator is 14 percent larger than its pole-to-pole diameter.
19H50.8M, +08°52' See pp. 26, 209

ANTARES: Visual magnitude 1.1. A class M supergiant, about 800 times the diameter of our sun.
16H29.4M, -26°26' See pp. 27, 41, 209

COAT HANGER: An asterism (random superposition of stars) reminiscent of a coat hanger.
BROCCHI'S CLUSTER OR COLLINDER 399
19H25.4M, +20°11'

CRESCENT: An emission nebula.
NGC 6888
20H12.0M, +38°21' See p. 221

DUMBBELL: A bright planetary nebula about 1,360 light-years away.
M27
20H59.6M, +22°43' See pp. 57, 132, 216, 219

EAGLE: Open cluster of stars in a diffuse emission nebula, about 7,000 light-years away.
M16
18H18.8M, -13°49' See pp. 15, 216, 219

HERCULES CLUSTER: Best globular cluster in northern hemisphere, about 25,000 light-years away.
M13
16H41.7M, +36°28' See pp. 216, 218, 220

LAGOON: A giant, bright interstellar cloud, emission nebula, and star cluster.
M8
18H03.8M, -24°23'

M5: Globular cluster. Distance is 24,500 light-years. Very old cluster with age of 13 billion years.
15H19.0M, +02°05'

RHO OPHIUCHI: Triple star embedded in a large blue reflection nebula.
IC 4604
16H25.6M, -23°28'

RING: One of the brightest planetary nebulae visible from the northern hemisphere.
M57
18H53.6M, +33°02' See pp. 216, 222

SEYFERT'S SEXTET: Misnumbered group of four foreground galaxies and one background galaxy.
NGC 6027
15H59.2M, +20°45'

SGR A: Black hole in Sagittarius at the galactic center. Not directly observable.
17H45.7M, -29°00'

SWAN: An emission nebula glowing like a neon sign.
M17 OR OMEGA NEBULA
18H20.8M, -16°11'

SNAKE NEBULA: A dark S-shaped dust lane obscuring the dense star clouds of the Milky Way.
B72 OR S NEBULA
17H23.5M, -23°38'

TRIFID: A three-petaled emission nebula.
M20
18H02.3M, -23°02'

VEGA: The fifth brightest star in the night sky at visual magnitude 0.0. Distance from Earth is 25.3 light-years.
18H36.9M, +38°47' See pp. 41, 55, 210, 214

VEIL NEBULA: A remnant from a supernova explosion 5,000 to 8,000 years ago. Entire loop spans 3°.
NGC 6992
20H57.1M, +31°13' See p. 221

WILD DUCK CLUSTER: A rich and compact open cluster containing about 2,900 stars.
M11
18H51.1M, -06°16'

Map Key ● Galaxy   -●- Planetary Nebula   ● Nebula (other than planetary)   ✣ Globular Cluster   ○ Open Cluster

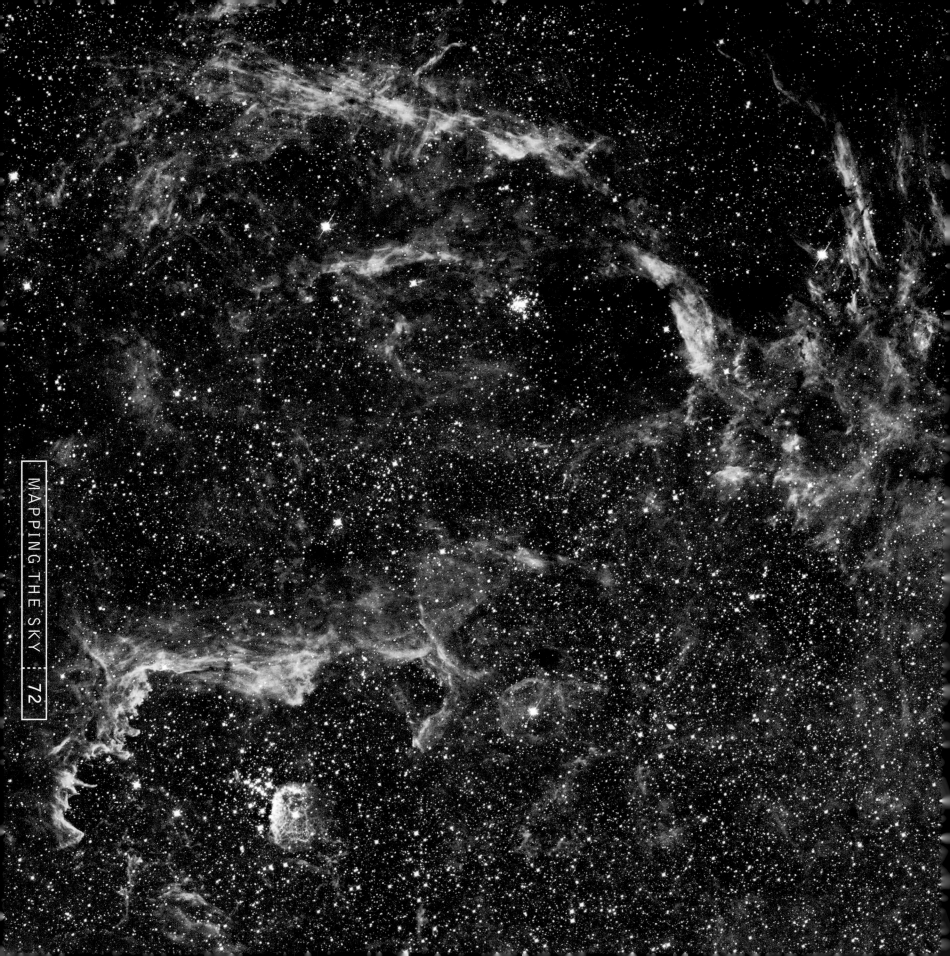

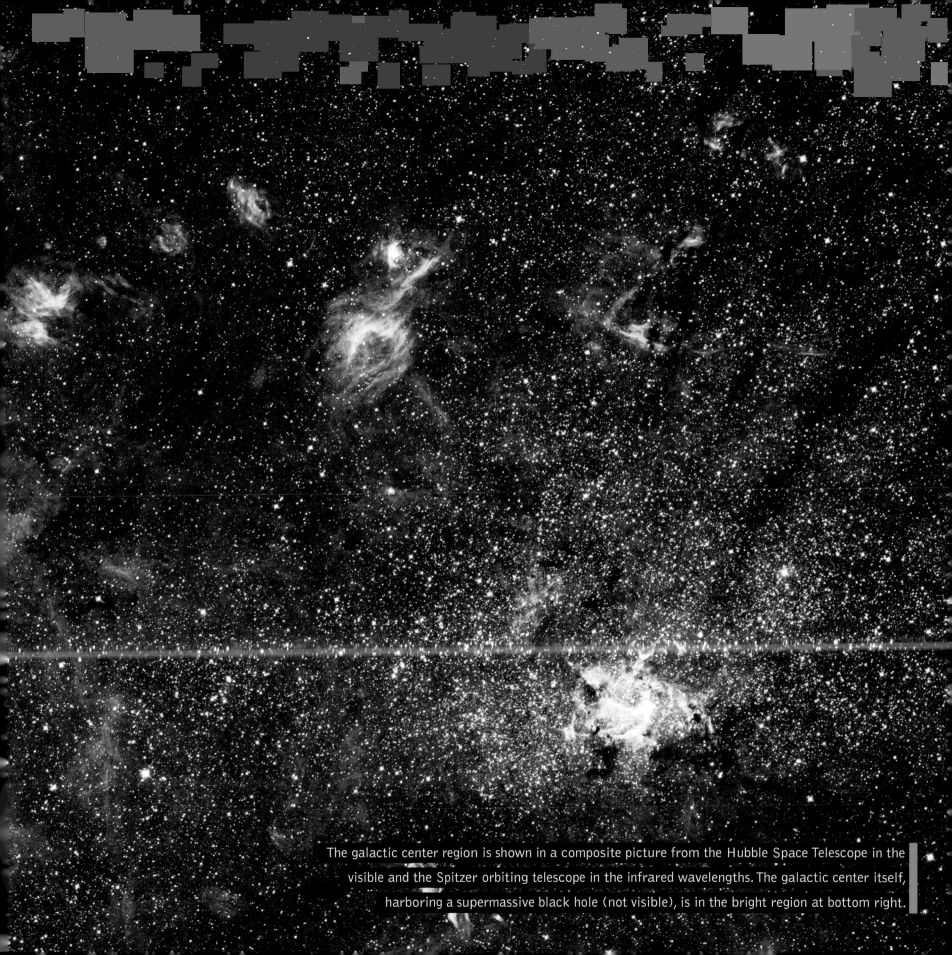

The galactic center region is shown in a composite picture from the Hubble Space Telescope in the visible and the Spitzer orbiting telescope in the infrared wavelengths. The galactic center itself, harboring a supermassive black hole (not visible), is in the bright region at bottom right.

Autumn

GO OUT AT MIDNIGHT ON SEPTEMBER 21 from mid-northern latitudes and you will see these stars above the southern horizon. The ecliptic crosses the celestial equator again on this chart. Two different great circles on a sphere always intersect twice. The great square of PEGASUS is prominent and the ANDROMEDA GALAXY is found nearby.

ANDROMEDA: Closest spiral galaxy to the Milky Way. The full angular diameter is more than 3°.
M31
00ʜ42.7ᴍ, +41°16' See pp. 36-37, 224-225

M15: Globular cluster. At 13.2 billion years old, one of the oldest clusters known.
21ʜ30.0ᴍ, +12°10'

BLUE SNOWBALL: Planetary nebula
NGC 7662
23ʜ25.9ᴍ, +42°35'

M74: Beautiful face-on spiral galaxy.
01ʜ36.7ᴍ, +15°47' See pp. 76, 134

CARTWHEEL: A ring galaxy showing a splash of star formation produced by collision with a companion galaxy.
00ʜ37.7ᴍ, -33°43'

M33: Spiral galaxy. Third largest in the Local Group.
01ʜ33.9ᴍ, +30°40' See pp. 134, 228, 231

EINSTEIN'S CROSS: Light from a background quasar is bent gravitationally around a foreground galaxy to produce four images.
PGC 69457
22ʜ40.5ᴍ, +03°21'

SCULPTOR GALAXY: Bright galaxy undergoing a period of intense star formation.
NGC 253
00ʜ47.5ᴍ, -25°17'

FOMALHAUT: A young star 2.3 times more massive than the sun. Has a toroidally (or donut) shaped protoplanetary dust disk.
22ʜ57.6ᴍ, -29°37' See p. 215

STEPHAN'S QUINTET: Visual grouping of five galaxies, four of which constitute a compact galaxy group—the first ever discovered.
22ʜ36.0ᴍ, +33°58'

HELIX: One of the closest planetary nebulae to Earth—only 650 light-years away.
NGC 7293 ᴏʀ Eʏᴇ ᴏꜰ Gᴏᴅ
22ʜ29.6ᴍ, -20°50'

M2: Globular cluster.
21ʜ33.5ᴍ, -00°49'

Map Key   Galaxy   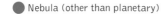 Planetary Nebula   Nebula (other than planetary)   Globular Cluster   Open Cluster

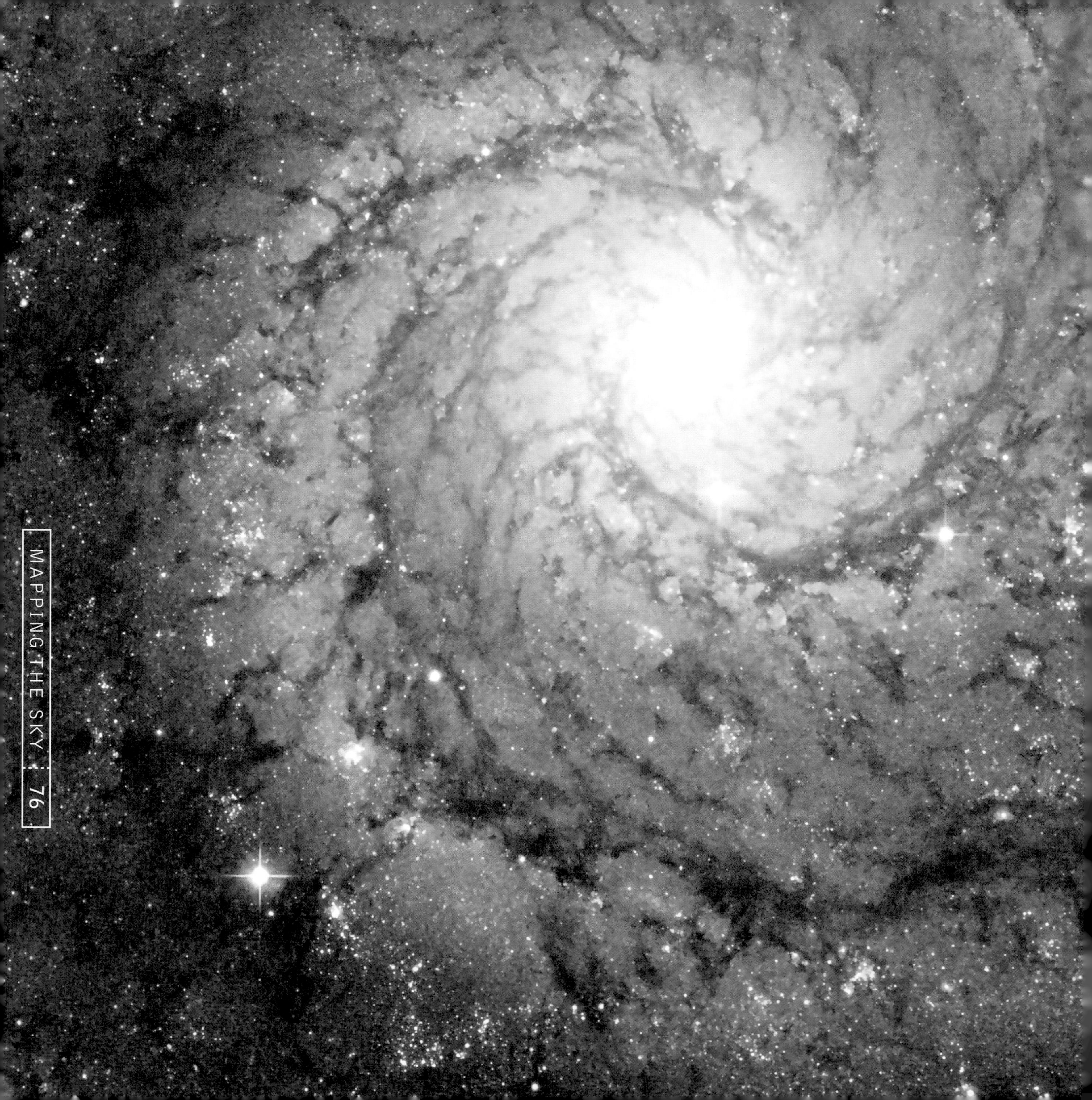

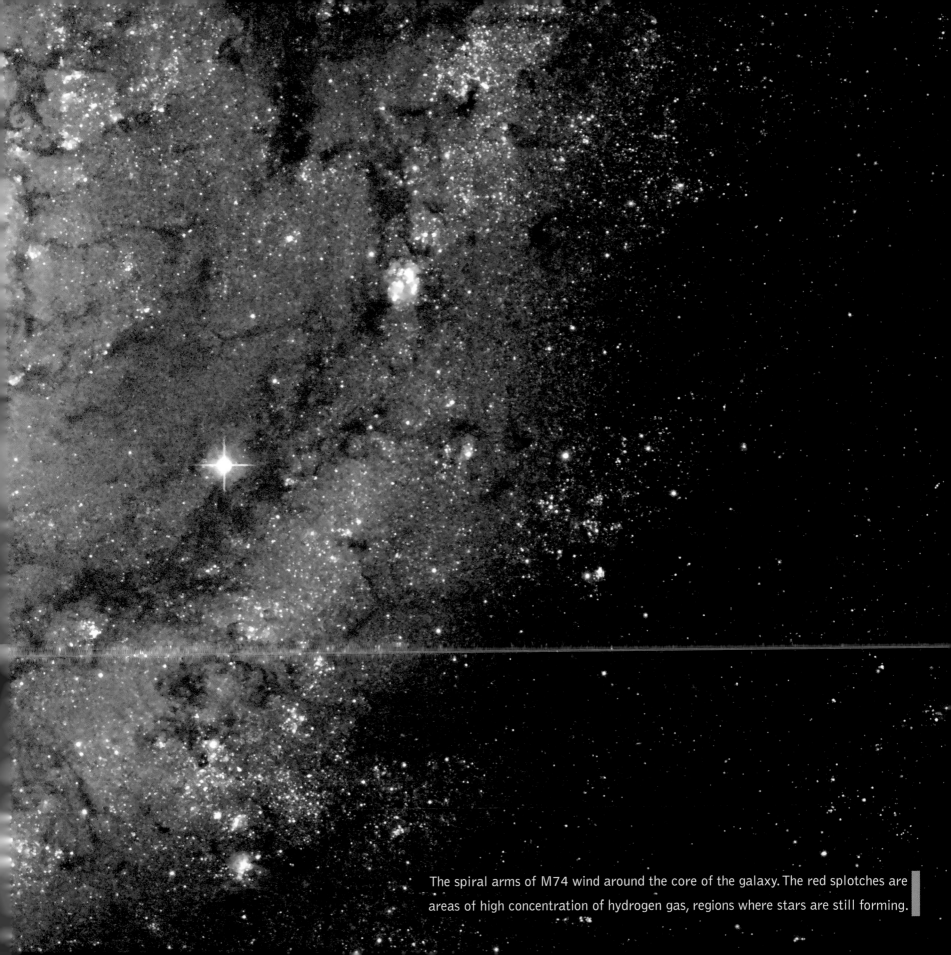

The spiral arms of M74 wind around the core of the galaxy. The red splotches are areas of high concentration of hydrogen gas, regions where stars are still forming.

South Circumpolar

THE SOUTH CIRCUMPOLAR STARS are not visible from mid-northern latitudes. From mid-southern latitudes (45° south of the Equator like New Zealand) these stars will always be above the *southern* horizon. The winter chart will still be visible at midnight on December 21 above the *northern* horizon, but upside down—because you are upside down!

ACHERNAR: Eighth brightest star in the night sky. Spins rapidly. Diameter at its equator is 50 percent larger than its pole-to-pole diameter.
01H37.7M, -57°14' See p. 54

ALPHA CENTAURI: Three-star system. Brightest is magnitude -0.3. Third component, Proxima Centauri, is our sun's closest neighbor.
14H39.6M, -60°50' See pp. 42, 55, 98

ACRUX: Brightest star in the Southern Cross.
ALPHA CRUCIS
12H26.6M, -63°06'

AGENA: The tenth brightest star in the night sky.
BETA CENTAURI
14H03.8M, -60°22'

ANT: A bipolar planetary nebula.
Mz3
15H17.2M, -51°59'

BECRUX: A spectroscopic binary star. (The stars are too close to separate visually.)
BETA CRUCIS OR MIMOSA
12H47.7M, -59°41'

CANOPUS: Second brightest star in the night sky. Primary navigation star for the Apollo astronauts.
06H24.0M, -52°42' See pp. 27, 55, 86

CIRCINUS GALAXY: A galaxy with an active nucleus, 13 million light-years away.
14H13.1M, -65°20'

COAL SACK: The most conspicuous dark nebula (its dust absorbs the background starlight).
12H51.0M, -63°00'

ETA CARINAE: Giant star four million times more luminous than the sun. Almost went supernova in 1843. Might blow up soon.
10H45.1M, -59°41'

GUM NEBULA: An emission nebula believed to be a supernova remnant from an explosion that took place millions of years ago.
08H30.0M, -45°00'

JEWEL BOX: An open cluster.
NGC 4755
12H53.7M, -60°22'

LARGE MAGELLANIC CLOUD: A dwarf galaxy in orbit around the Milky Way. Fourth largest member of our Local Group of galaxies.
05H23.7M, -69°45' See pp. 24, 27, 134, 216

OMEGA CENTAURI: Globular star cluster.
NGC 5139
13H26.8M, -47°29'

SMALL MAGELLANIC CLOUD: A dwarf galaxy in orbit around the Milky Way.
NGC 292
00H52.7M, -72°50' See pp. 24, 27, 134

TARANTULA: Emission nebula in the Large Magellanic Cloud. Supernova 1987A occurred near here.
NGC 2070
05H38.6M, -69°06'

VELA SNR: Supernova remnant associated with the Vela pulsar (neutron star).
NGC 6955
08H35.3M, -45°11'

Map Key ● Galaxy ✦ Planetary Nebula ● Nebula (other than planetary) Globular Cluster Open Cluster

DISTA

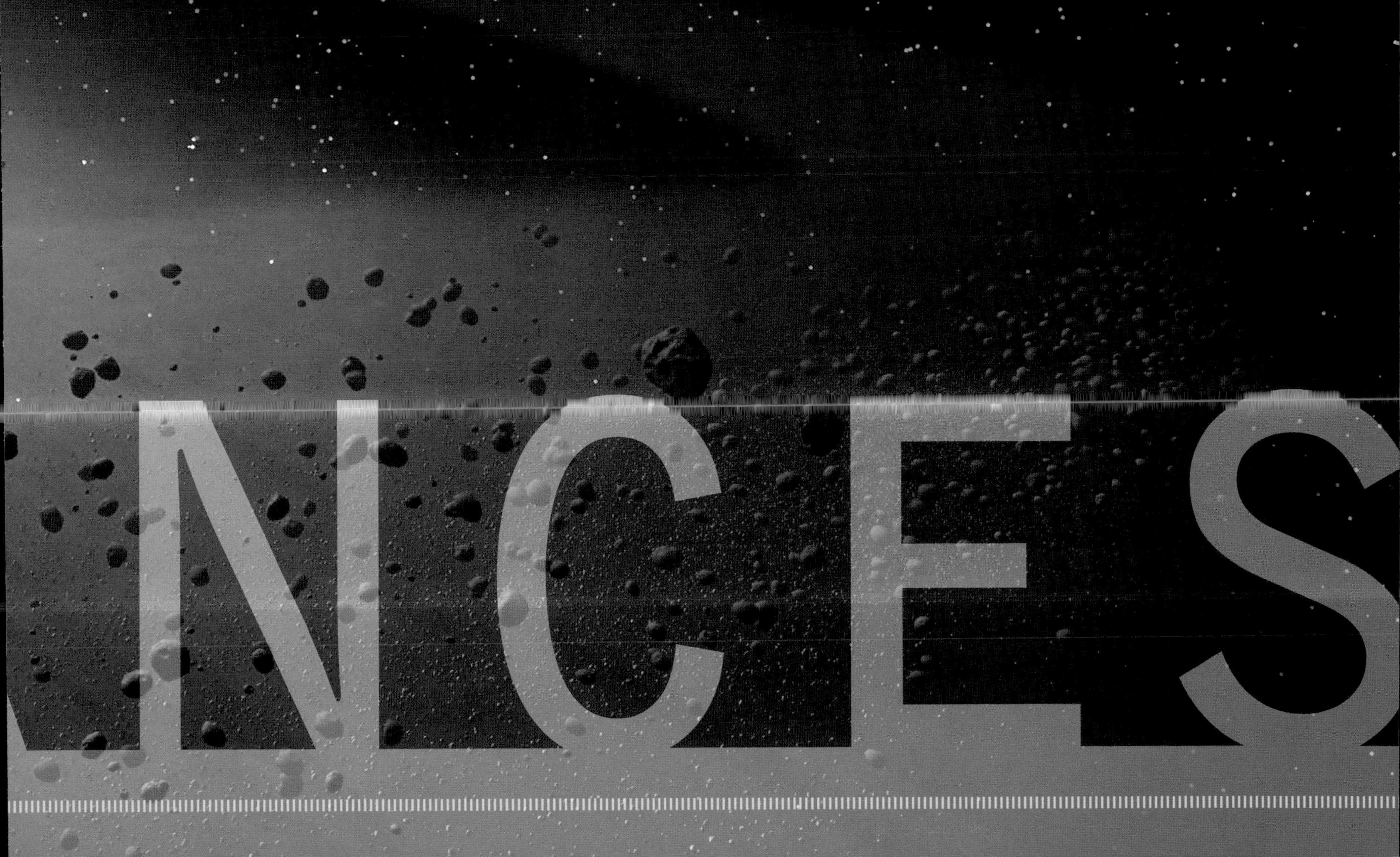
NCES

CHAPTER 3

IN THE BEGINNING

IF WE WANT TO KNOW the sizes of the objects we see in the sky, we need to know their angular diameters and how far away they are. The sun and moon both have the same angular diameters in the sky—one-half degree. That means that if they were at the same distance from Earth, they would be the same size. But since the sun is farther away, it is bigger. We now know that the sun is about 400 times farther away than the moon, and therefore 400 times as large as the moon.

We can figure out the physical size of an object if we know its angular size and its distance. We use the principles of Euclidean geometry. The circumference of a circle is π times its diameter; π is 3.14159265 . . . , a decimal number with an infinite number of digits following the decimal point. Trillions of digits of π have been computed, with more being added all the time. The radius of a circle is half its diameter, so the length of the circumference is just 2π times its radius. The circumference of a circle is 360 degrees.

Mathematicians call the angle (360 degrees) / 2π = 57.295 . . . degrees one radian. It can be used as a unit of measure. To convert from degrees to radians, just multiply by 2π / (360 degrees). Using radians, the rule for computing the physical size of a small object from its angular size and its distance is especially simple: For angular sizes smaller than a few degrees, the physical diameter of an object is equal to its distance times its angular diameter in radians. Thus we can determine the true physical sizes of objects if we know their angular diameters and their distances. How big are the objects we see in the sky? To find out, we need to know how far away they are. But how to determine those distances?

Ancient Egyptians and Babylonians charted the course of the stars in the heavens and measured the lengths of the seasons. Stonehenge in England, built about 1900 B.C., was carefully oriented so that on the longest day of the year the sun rose over its Heel Stone. But it was the ancient Greeks who first began to figure out what was actually going on in the heavens—and how far away the moon and sun might be from Earth.

Pythagoras (582–497 B.C.) was one of the first to teach that Earth is a sphere. Aristotle (384–322 B.C.) adopted and reinforced Pythagoras's view. The surface of Earth is curved: Ships disappear over the horizon hull first, mast

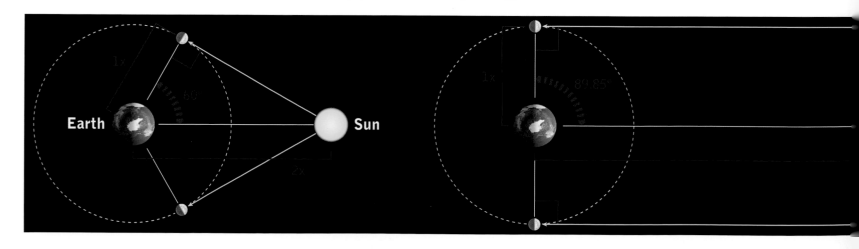

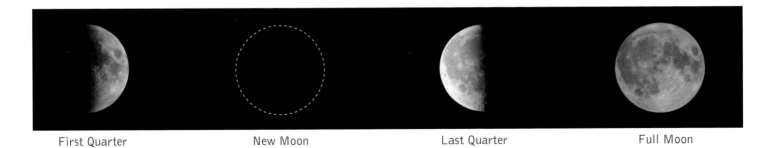

First Quarter   New Moon   Last Quarter   Full Moon

last. Also, as one travels farther and farther south, new stars appear above the horizon that were not visible in the north. That would not occur if Earth were flat. Finally, during a lunar eclipse, when Earth passes directly between the sun and the moon, you can see Earth's shadow on the moon. Its outline is circular—always. No matter where the moon is relative to your observation point, the shadow of Earth always appears to be circular. The only shape that always casts a circular shadow is a sphere.

So people have known that Earth is a sphere since the time of Aristotle and before. However, they still assumed that Earth was at the center of the universe. Eudoxus of Cnidus (408–355 B.C.) thought the celestial sphere carrying the stars surrounded a stationary, spherical Earth at the center. The celestial sphere rotated on its axis once a day, causing the stars to rise and set. A transparent sphere nested inside the celestial sphere carried the sun on a 360-degree journey relative to the stars once a year. The moon and planets were supposed to move on a complicated additional set of nested spheres, with various rotations. Aristarchus of Samos (320–250 B.C.) disagreed—he said the celestial sphere does not move. Instead, it is Earth that rotates on its axis once a day. The stars rise and set, appearing to move, simply because Earth is turning. He was right!

The moon orbits Earth once a month, and Aristarchus figured out that it shines by the light it receives from the

Phases of the moon (above). Where a quarter moon would be located if the sun were not much farther away than the moon (below left) and if the sun were much farther away than the moon—which is actually the case (below right).

Sun →

sun. At first quarter, the moon is half illuminated by the sun as seen from Earth (previous page). First quarter moon occurs when a right angle is formed between Earth and the sun, as seen from the moon. The moon continues to circle in its orbit until last quarter is reached. Again, the angle between Earth and sun, as seen from the moon, is a right angle, and the moon appears half illuminated as seen from Earth.

Aristarchus noticed that if he could measure the angle between the moon and the sun as seen from Earth at first quarter, he could determine the shape of the triangle formed by Earth, moon, and sun at that time and figure out how much farther away the sun was, compared with the moon. In the illustration on page 82, we have placed the sun just twice as far away as the moon. In that case, the angle between the sun and moon as seen from Earth at first quarter would be 60 degrees. If we move the sun farther away, this angle gets larger, approaching 90 degrees as we keep moving the sun farther and farther away.

Aristarchus measured the angle in the sky between the moon and the sun when the moon was at first quarter and found it to be 87 degrees. From this triangle shape, he deduced that the sun was about 19 times farther away than the moon. Since the sun and moon have approximately the same angular diameter in the sky, Aristarchus reasoned that the sun, being 19 times farther away, must be 19 times as large as the moon. (The actual angle between the sun and the moon in the sky at first quarter is 89.85 degrees, not the 87 degrees Aristarchus measured. So, this puts the sun about 400 times farther away than the moon and makes it 400 times larger than the moon, but Aristarchus was on the right track.)

The moon is in a nearly circular orbit around Earth and circles it at a nearly constant speed. If the sun is far away (previous pages, right; where it is far beyond the right edge of the figure), then rays from the sun illuminating the moon come in as virtually parallel rays, and first quarter and last quarter are equally spaced on opposite sides of the moon's orbit. This picture matches your own experience—you already know that the four quarters of the moon (first quarter, full, last quarter, new) are approximately equally spaced around the month, being about a week apart. Because the phases of the moon are equally spaced around the orbit, it means that the sun must be very far away relative to the moon.

Judging from the size of Earth's shadow relative to the moon during a total lunar eclipse (opposite), Aristarchus deduced that Earth is about three times larger in diameter than the moon. Aristarchus figured that if the sun is 19 times larger than the moon, and Earth is 3 times larger than the moon, that would mean the sun is 19/3 (or 6.67) times larger than Earth! The sun is larger than Earth. If the smaller moon orbited the larger Earth, Aristarchus said it made sense for Earth to orbit the still larger sun. Why should the enormous sun orbit the tiny Earth? Aristarchus boldly concluded that Earth orbits the sun. The sun marches in a circle around the celestial sphere once a year simply because Earth circles the sun once a year, and the sun's position against the background stars rotates accordingly. Earth is not at the center of the universe. In this conclusion, he anticipated Copernicus by more than 1,700 years.

Total lunar eclipse. By looking at the "bite" taken out of the moon by Earth's shadow as the moon moves into it (right to left in this sequence and diagrammed beneath), Aristotle deduced the shape of Earth. Aristarchus deduced its size (about three times that of the moon). Vanderbei took this sequence of images on November 11, 2003, with a 3.5-inch Maksutov-Cassegrain (reflector) telescope and a Starlight Express MX9 CCD camera.

Earth's
Shadow

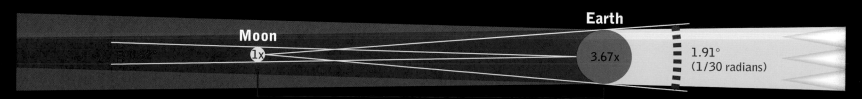

Moon

Earth

1x

3.67x

1.91°
(1/30 radians)

30x Earth's Diameter

Measuring the Earth

SO HOW BIG IS THE EARTH? Eratosthenes (276–194 B.C.) found out. He worked in Alexandria, Egypt, where he was the chief librarian of the Library of Alexandria. He knew that at noon on the longest day of the year, June 21, the sun famously stood directly overhead in the town of Syene, south of him on the Nile River (at modern-day Aswan), because one could look down a deep well there at that time and see the reflection of the sun in the water.

But he also knew that the sun did not stand directly overhead at noon on June 21 in Alexandria. An obelisk in Alexandria cast a shadow then. If the sun were directly overhead, it would cast no shadow. By measuring the height of the obelisk and the length of the shadow, he determined that the sun was 7.2 degrees off vertical at noon on June 21 in Alexandria.

The difference was the result of the curvature of Earth. Eratosthenes knew the sun was very far away, so light rays from the sun essentially traveled on parallel lines toward Earth. If Earth's surface curved by 7.2 degrees between Syene and Alexandria, as it would if they were separated by 7.2 degrees of latitude on the globe, then that would cause the sun to be 7.2 degrees off vertical in Alexandria, as shown opposite. The angle between Alexandria and Syene as seen from the center of Earth should be equal to the amount by which the sun was off vertical in Alexandria (the two red arcs). These angles are equal if the rays from the distant sun are essentially parallel, according to a famous theorem in Euclidean geometry.

Because 7.2 degrees is one-fiftieth of a full circle (360 degrees), Eratosthenes reasoned that the distance between Syene and Alexandria should be one-fiftieth of the circumference of Earth. Now all he needed to do was send someone on a trip directly south from Alexandria to Syene to measure the distance between them. Multiply this number by 50, and he would have the circumference of Earth. He got a circumference of about 40,000 kilometers (25,000 miles) in today's units, pretty close to the true value.

When you divide this number by π, you get the diameter of Earth: approximately 12,730 kilometers (7,910 miles). Eratosthenes' method was important because it set the scale for distances measured later, which would often be measured first relative to the size of Earth.

Later, Posidonius (135–50 B.C.) repeated Eratosthenes' work, using the star Canopus. He measured in degrees how far it rose above the southern horizon as seen from the Greek island of Rhodes, compared with Alexandria, which was farther south. For Earth's circumference, Posidonius obtained a value of only about 29,000 kilometers (18,000 miles)—quite a bit too small. However, Posidonius was friends with Cicero and Pompey, and the influential second-century astronomer Ptolemy chose Posidonius's value over Eratosthenes'.

That wrong answer of Posidonius affected the course of world history. People learned from Marco Polo's travel accounts the approximate overland distance between Europe and China. Using Posidonius's value for the circumference of Earth, Christopher Columbus figured the distance he would have to sail, traveling west from Europe, to circle the globe and arrive in China. With Posidonius's smaller Earth, the trip looked short and favorable for a trade route—so he embarked.

Eratosthenes' method for measuring the diameter of Earth using observations from Alexandria and Syene

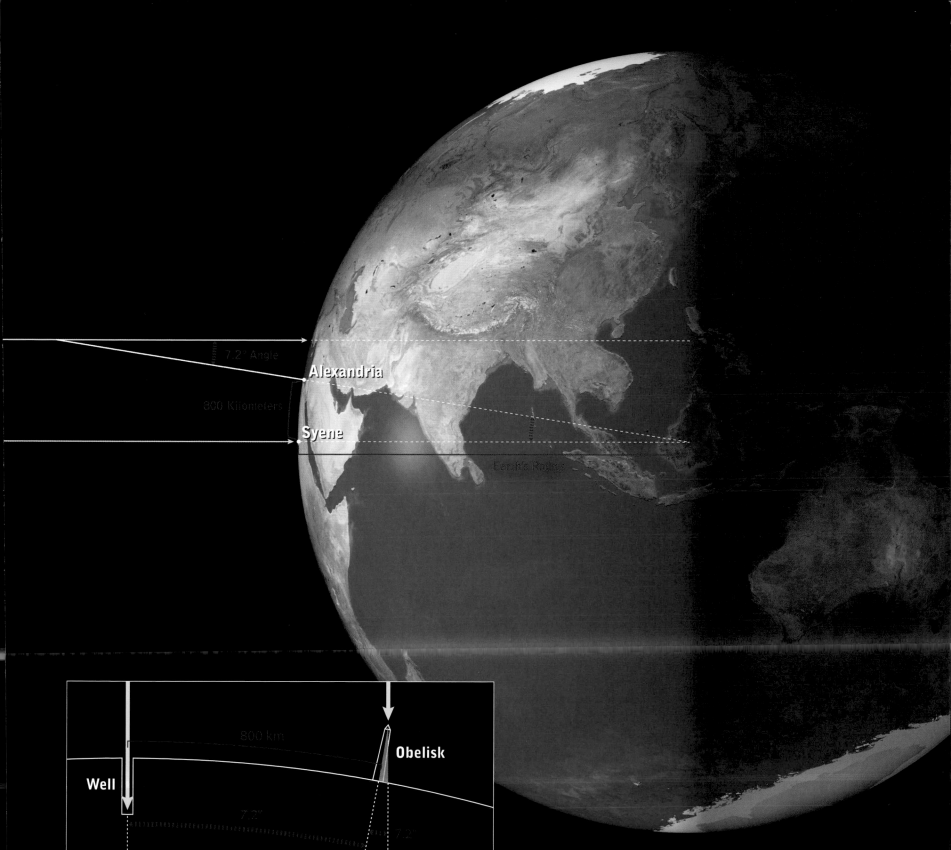

FROM THE EARTH TO THE MOON

REFINING ARISTARCHUS'S METHODS, Hipparchus (190–120 B.C.) determined accurate values for the size of the moon and its distance from Earth relative to the diameter of Earth. Hipparchus knew the correct angular diameter for the moon (he used 0.55 degrees—quite close to the actual value of 0.5181 degrees). He measured the size of Earth's shadow during a lunar eclipse and found it to be 2.5 times the diameter of the moon. Using a geometrical argument similar to Aristarchus's, he concluded that Earth was about 3.5 times as large as the moon. (Aristarchus's calculations of the relative sizes of Earth, the moon, and the sun had been quite sophisticated. He knew that if the sun is larger than Earth and very far away, the shadow of Earth in space would trace out a cone that got narrower the farther away from Earth you got. He figured correctly that in this case, the size of Earth's shadow on the moon would fall short of Earth's true size by about one lunar diameter.)

If Earth is 3.5 times bigger than the moon, Hipparchus reasoned that if it were viewed from the same distance as the moon, it would look 3.5 times bigger, or about 1.75 degrees across. Now, 1.75 degrees is about one-thirtieth of a radian (which is approximately 57 degrees), so that means that the distance to the moon is about 30 Earth diameters (below). Usually we determine an object's distance and use it with its angular size to tell us its physical diameter. Here the comparison with Earth's shadow tells us the physical size of the moon relative to Earth, which combines with its angular size to tell us its distance in Earth diameters.

FYI Hipparchus figured out that the distance to the moon is about 30 Earth diameters— probably farther away than you thought.

Hipparchus had a second way to check his figure for the distance to the moon. There was a total eclipse of the sun observed near the Hellespont (a strait in present-day Turkey) in 190 B.C. At that location, the moon exactly covered the sun. But Hipparchus knew that in Alexandria, Egypt, on the same day, the moon covered only four-fifths of the sun, making a partial eclipse. The moon

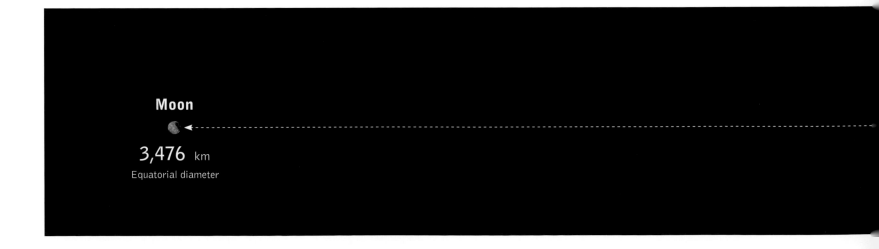

Moon

3,476 km
Equatorial diameter

moved one-fifth of its diameter relative to the distant sun as seen from the two locations. This is called the parallax effect, a term that will recur often.

Parallax is important to us every day. Your brain uses this effect from your two eyes to achieve stereo vision and enable you to judge distances. Point your index finger at a distant object on the horizon. Now close one eye and then the other. From one eye, your finger will still point at the designated object. That is your dominant eye. From the other eye, your finger will be pointing about 2.5 inches off target. Blink one eye and then the other and you will see your finger jump back and forth those 2.5 inches. Your finger appears at different locations against the distant background from each eye, jumping those 2.5 inches because that is the distance between your two eyes. When your brain fuses these two images, it uses those differences to tell you the 3-D distance to your finger.

Determining the distance to the moon using the difference between those two views of it is like having one eye in the Hellespont and the other in Alexandria and getting a stereo view. The sideways shift between your view from the two locations must be equal to one-fifth the diameter of the moon—the amount it jumps between the two views.

Hipparchus knew the latitudes of the Hellespont (40°) and Alexandria (31°). The latitudes could be determined by knowing what stars in the celestial sphere passed directly overhead at the two locations. He also knew the direction of the line of sight to the sun from the two locations. That allowed him to calculate the sideways shift between the two lines of sight in terms of the diameter of Earth. This information, together with the known angular diameter of the moon, enabled him to calculate the distance to the moon in terms of the diameter of Earth. Consistent with the shadow method, he found the moon to be roughly 30 Earth diameters away. But to get an accurate distance to the moon as shown in the diagram below, it is necessary to know the actual diameter of the Earth—which Eratosthenes provided a generation before.

The relationship between Earth and the moon in space. The sizes of each object and distance between them are shown to scale.

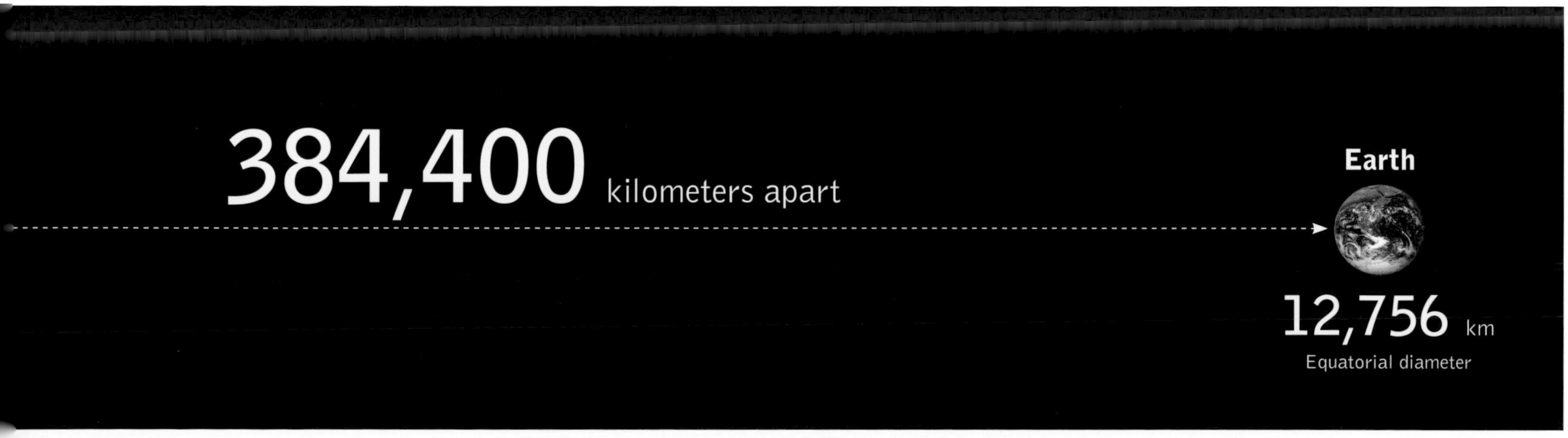

384,400 kilometers apart

Earth

12,756 km
Equatorial diameter

To the Sun

THE DISTANCE BETWEEN Earth and the sun is called the astronomical unit, or AU for short. How do we find it?

First we find the relative sizes of the orbits of the planets. Venus's orbit is nearly circular, and as it and Earth both circle the sun at different rates, Venus keeps lapping Earth, oscillating from one side of the sun to the other as time progresses (opposite). The farthest Venus ever gets from the sun is 46 degrees, shown by its two limiting positions at the top and bottom. If you draw dots for the sun and Earth, you can draw with a protractor two lines at 46 degrees with respect to the sun, just like the diagonal lines shown in the figure. Now draw a circle around the sun with a compass that just touches those two diagonal lines. That's the circular orbit of Venus drawn to the same scale as the distance between Earth and the sun. If Venus wanders to a maximum elongation of 46 degrees with respect to the sun, the radius of Venus's orbit is 0.72 AU. After Venus reaches greatest eastern elongation from the sun as the evening star, the angle between it and the sun begins to drop below 46 degrees. We can draw the angle we observe and extend a line from Earth at that angle. Where it intersects the orbit of Venus, that's Venus's position. Measuring the length of this line segment and comparing it with the distance between Earth and the sun allows us to calculate the distance to the sun in terms of the distance between us and Venus at a given time. We can find the distance between us and Venus using the parallax effect by observing from two locations as shown below.

Occasionally Venus passes in front of the bright star Regulus (an occultation). Suppose we observe from South

Orbits of Earth and Venus (opposite). Seen from South Africa (below), Venus passes directly in front of the bright star Regulus; from England, it misses. We can measure the red angle at England—it's equal to the red angle at Venus. That tells us the angular separation of England and South Africa (a known distance apart) as seen from Venus, and therefore the distance to Venus.

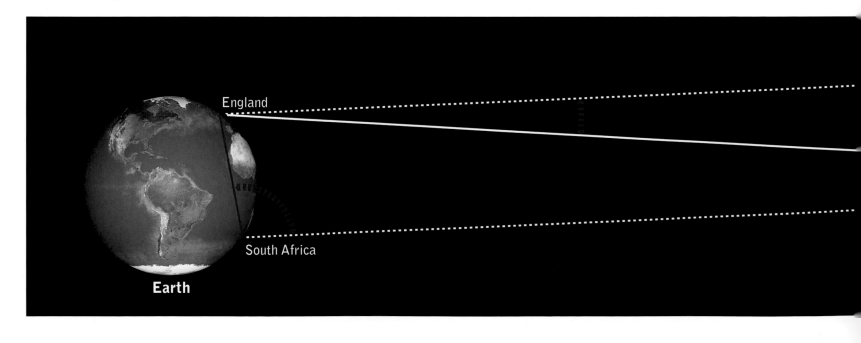

England

South Africa

Earth

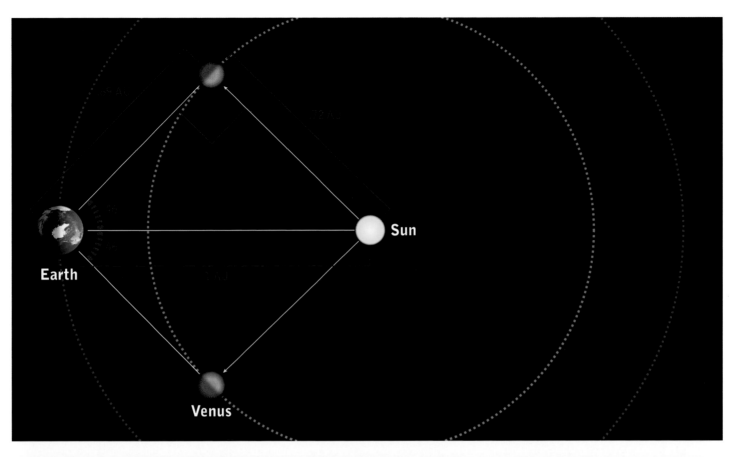

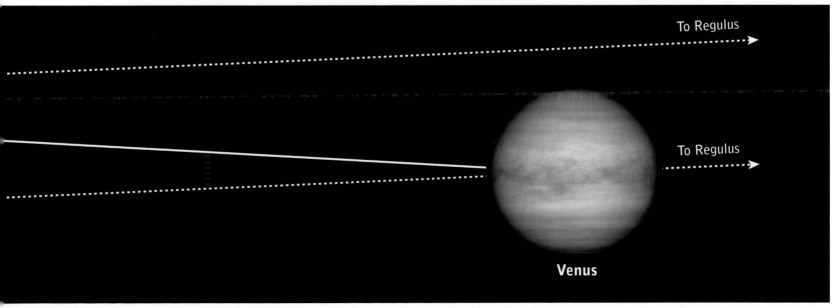

Africa and England, and from South Africa we see Regulus pass behind Venus on a line that would take it right behind the center of Venus. Regulus is so far away that lines of sight to it are essentially parallel.

The line of sight to Regulus from England misses Venus entirely. Observers in England will see no occultation at all. The angle between Regulus and the center of Venus can be measured by an observer in England. The angle between Regulus and the center of Venus as seen from England is the same as the angle between England and South Africa as seen from Venus. This is true according to the same theorem from Euclidean geometry about parallel lines that Eratosthenes used earlier. We know the distance between the two observatories, so if we know their angular separation as seen from Venus, we can determine the distance between Venus and Earth at that moment.

FYI One astronomical unit, or AU, is defined as the mean distance between Earth and the sun— 149,600,000 kilometers. Jupiter's distance to the sun is 5.2 AUs—that is, 5.2 times the Earth's.

This example illustrates the technique. But we don't actually need to have an occultation to use it—near misses can be used as well. The same trick can be used to find the distance to Mars using a distant background star passing close by. Knowing the position of Mars on its orbit, that distance can then be used to figure the distance to the sun. Applying this parallax method to observations of Mars taken in 1672 in Paris and Cayenne, in modern-day French Guiana, South America, Giovanni Cassini (1625–1712) estimated the distance to the sun to be 140 million kilometers (87 million miles), close to the true value. One can also use this method with an asteroid. The asteroid's orbit is an ellipse, and its movement around the ellipse is not uniform, but predictable. It's just a question of doing a little more math to find the answer.

Another method to establish the length of the astronomical unit uses transits of Venus, when Venus passes directly between Earth and the sun. These events can be viewed from two widely separated locations (north and south) on Earth. Venus's path crossing the sun will be seen slightly differently from the two locations. The northern observer will see Venus crossing the sun on a path that appears farther south; therefore, the transit will have a different duration from start to finish than that timed by the other observer. Unfortunately, attempts to use this method in 1761 and 1769 failed to produce an accurate measurement. Captain James Cook observed the transit from Tahiti in 1769.

Today, of course, we can measure the distance between us and Mars to determine the astronomical unit with high accuracy simply by measuring the speed of light in the lab and then measuring the time delay to send and receive radio waves (which also travel at the speed of light) from space probes sitting on the surface of Mars. Earth's orbit is approximately, but not exactly, circular. An astronomical unit, defined as the mean distance between Earth and the sun, is 149,600,000 kilometers (93,000,000 miles). Given the distance of the sun and its angular size we can calculate the sun's diameter: 1,392,000 kilometers (864,949 miles). Aristarchus was right in thinking the sun was much bigger than Earth and that our planet goes around it.

Gott and Vanderbei observed and photographed this transit of Venus on June 8, 2004. Venus can be seen as a dark dot in silhouette against the bright face of the sun. Its next transits won't be until June 5–6, 2012, and December 10–11, 2117.

To the Stars

SO WHY DIDN'T PEOPLE BELIEVE Aristarchus? Because Aristotle had a killer argument for why Earth did not move. If Earth circled the sun, the stars should show a parallax effect—and this was not seen. As Earth circled the sun, Earth's position relative to the stars should oscillate, causing the stars' positions to oscillate once a year in the sky. This is explained in the figure opposite. The true situation is as shown at the top—just as Aristarchus envisioned it. Earth circles the sun once a year. Assume the stars and the sun remain at fixed positions. How does it look from Earth? We are riding on Earth, so it looks to us like Earth is not moving.

It looks to us like the sun moves in a small circle of radius 1 AU around Earth once a year (that's why it circles the celestial sphere once a year). The stars do not move relative to the sun, so as seen from Earth, stars must, like the sun, also seem to move in 1 AU circles over the course of a year. We should be able to see the stars trace these circles in the sky every year. These parallax circles represent the reflex motion of the stars relative to Earth produced by the motion of Earth as it circles the sun, creating changing viewing angles during the year to those stars (top right). If the distance from Earth to the sun is 1 AU, then the radius of all these parallax circles would also be 1 AU. The angular radius of the parallax circle depends on the distance to the star. A nearby star has a larger angular oscillation in the sky as seen from Earth than a distant star (bottom right).

If we look at a constellation, the nearby stars should oscillate more during the year than the distant stars. So the positions of nearby stars should shift during the year relative to more distant stars. The ancients thought that the stars were close enough that these oscillations should have been visible to the naked eye. But none were seen. Aristotle thought that proved Earth didn't move.

Aristarchus proposed an answer—no parallax effects were seen because the stars were infinitely far away. Parallax effects get smaller the farther away the stars are. Put the stars twice as far away, and the parallax effects become half as large. Put them infinitely far away, and the parallax effects disappear entirely. It was almost the right answer.

In 1453 Nicolaus Copernicus (1473–1543) published a sun-centered model based on Aristarchus's work. In it he was able to explain in a simple manner the main motions seen in the solar system. Mercury and Venus oscillate back and forth ahead of and behind the sun as the sun circles the sky once a year. Copernicus said this is because they, like Earth, orbit the sun but are closer to the sun than Earth is.

Before Copernicus, people had explained this motion with epicycles: The planet was supposed to circle a point that itself circled Earth. The big circle carrying the point was called the deferent, and the small circle around that point was called the epicycle. Venus and Mercury had large deferent circles exactly synchronized with the sun. Their epicycles produced their oscillations around the sun. The outer planets (Mars, Jupiter, Saturn) had big deferent circles that traced their slow orbits around the sky and epicycles with periods of one year each, which in reality showed the reflex (parallax) motion relative to Earth caused by Earth's movement around the sun. The

If we circle the sun (top), it should appear to us as if all stars execute circular motions. Observe a star on June 1 and February 1; it is in different angular positions in the sky (top). Relative to us (bottom) it seems to execute a circular motion. The diagram is not to scale—stars are much farther away.

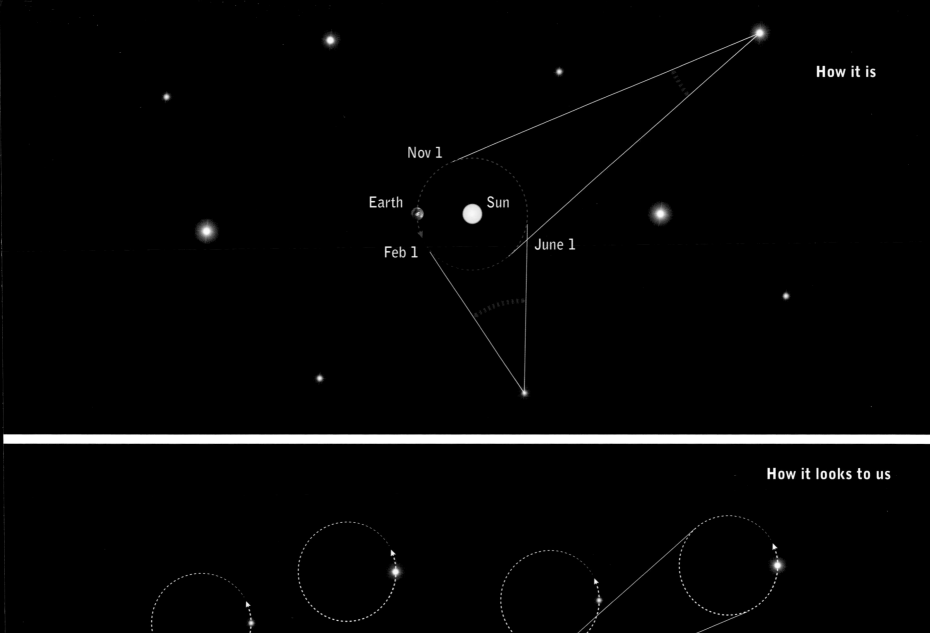

How it is

Nov 1

Earth   Sun

Feb 1   June 1

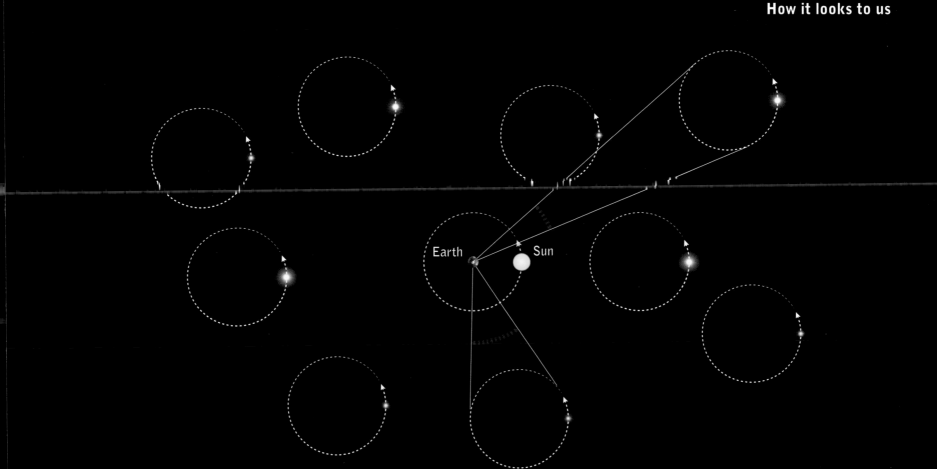

How it looks to us

Earth   Sun

planets were *already* showing the parallax oscillations, caused by the motion of Earth around the sun, that Aristotle was looking for!

Those parallax circles—epicycles for the outer planets, and deferent circles for Mercury and Venus—all had a period of one year. That should have been suspicious—all those synchronized deferent and epicycle circles with periods of exactly one year.

FYI A parsec is **30,856,780,000,000** kilometers, or **3.262** light-years. Proxima Centauri, the closest star (besides the sun), is **1.295** parsecs from Earth.

There are some small additional complications because the orbits of the planets are actually ellipses, not perfect circles, but Copernicus explained the major observed motions in a simple and compelling fashion. Of course, Copernicus still had to answer Aristotle's question of why the stars showed no naked-eye parallax. Copernicus's answer was that the stars were simply very far away (more than 1,000 AU), so the parallax circles were just too small to observe with the naked eye (a radius of oscillation of less than 3.5 minutes of arc). That was the right answer.

Nicholas of Cusa (1401–1464) proposed that the stars were other suns like our sun, and he was right. Christiaan Huygens (1629–1695) realized that if the stars were other suns, they must be very far away, because they are very faint compared with our sun. He figured that stars differed in brightness simply because they are at different distances from Earth. The brightest star, Sirius, must be the closest. So he set out to find out how far away Sirius is. He took a 12-foot-long tube and covered one end with a small plate in which he had punched a pinhole only 1 / 144 of an inch

in diameter. Viewed from the other end of the tube, the pinhole had an angular diameter 1 / 182 as wide as the sun. He then put a tiny glass ball in the hole to serve as a very short focus lens, which, when the tube was pointed at the sun, produced a tiny image of the sun that was only 1 / 152 as large as the pinhole itself. That meant when he looked at this image from the other end of the tube, his head shrouded to keep out the daylight, the image had an angular diameter only 1 / (182 x 152) = 1 / 27,664 as large as the sun does in the sky. The tiny image looked as bright to Huygens as Sirius did at night. So if Sirius was just like the sun, it must be 27,664 times farther away!

That supported Copernicus. Sirius was so far away that it would show no observable parallax to the naked eye. Huygens got the distance to Sirius right to within a factor of 20, quite an accomplishment. Sirius is actually somewhat larger and brighter than the sun, and therefore farther away than he figured. But Huygens was right—the stars were other suns.

Astronomers continued to look for the parallax motion of the stars. Friedrich Wilhelm Bessel (1784–1846) finally found it in 1838. He measured the distance to 61 Cygni, a dim star that is actually closer than Sirius, and found it to be 56 trillion kilometers (35 trillion miles). Soon parallax distances to many stars had been measured.

The standard unit for stellar distances used today is the parsec, a distance at which a star would show a parallax oscillation radius of one second of arc back and forth in the sky. A parsec is equal to 3.262 light-years. Sirius is 2.65 parsecs away. Since a light-year is the distance light can travel in a year, it takes light from Sirius 8.6 years to reach us.

Stars in the neighborhood of the sun (opposite). Accurate parallax measurements allow us to plot their positions.

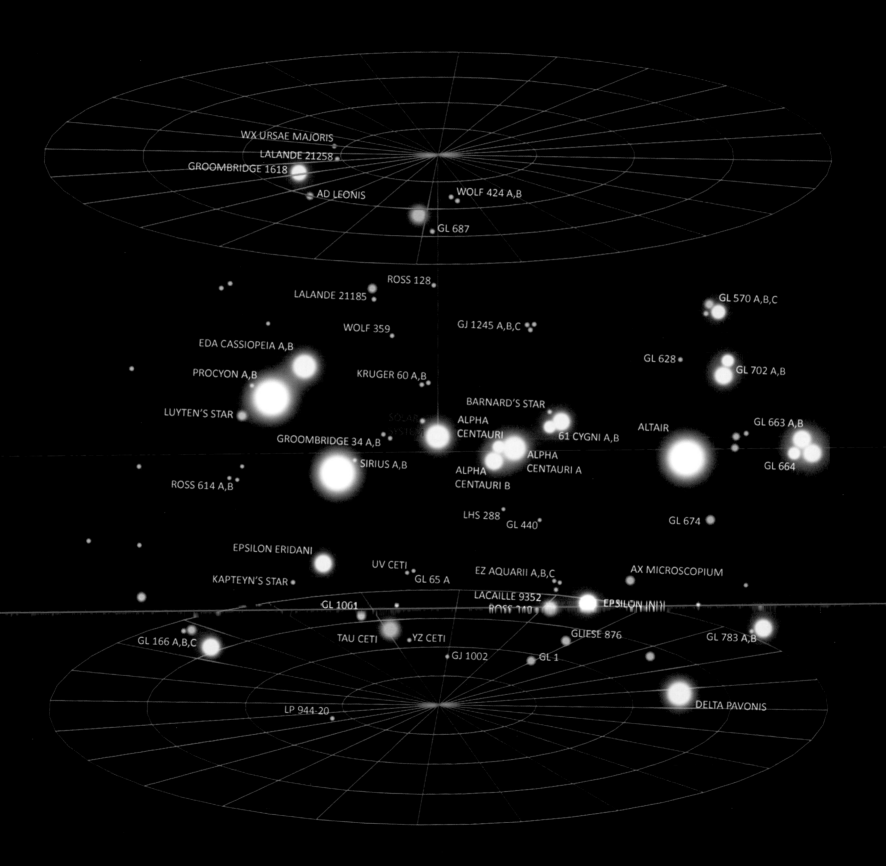

WX URSAE MAJORIS
LALANDE 21258
GROOMBRIDGE 1618
AD LEONIS
WOLF 424 A,B
GL 687

ROSS 128
LALANDE 21185
GL 570 A,B,C
WOLF 359
GJ 1245 A,B,C
EDA CASSIOPEIA A,B
GL 628
GL 702 A,B
PROCYON A,B
KRUGER 60 A,B
BARNARD'S STAR
LUYTEN'S STAR
SOLAR SYSTEM
ALPHA CENTAURI
61 CYGNI A,B
ALTAIR
GL 663 A,B
GROOMBRIDGE 34 A,B
ALPHA CENTAURI A
GL 664
SIRIUS A,B
ALPHA CENTAURI B
ROSS 614 A,B
LHS 288
GL 440
GL 674
EPSILON ERIDANI
UV CETI
EZ AQUARII A,B,C
AX MICROSCOPIUM
KAPTEYN'S STAR
GL 65 A
LACAILLE 9352
GL 1061
ROSS 110
EPSILON INDI
TAU CETI
YZ CETI
GLIESE 876
GL 783 A,B
GL 166 A,B,C
GJ 1002
GL 1
DELTA PAVONIS
LP 944-20

» STEREO VIEW «

OUR SOLAR NEIGHBORHOOD

THIS PAGE SHOWS TWO simulated pictures taken from two different, widely separated locations. If you compare them, you will see that some stars (the foreground stars) are shifted more than others from one picture to the next. This is parallax. We could calculate the distance to each star by measuring these shifts. Or you could let your brain do it—using

stereoscopic vision to see the view in 3-D.

Given that we sense depth in the real world when our two eyes see things from slightly different perspectives,

we can trick ourselves into seeing a 3-D scene even on the flat pages of a book—all we need to do is give two separate images, one from the point of view of the left eye and one from the point of view of the right eye, and arrange it so that the left eye looks at the left eye's image and the right eye looks at the right eye's image.

In this stereo pair of images, the left eye's image is the one on the right and the right eye's image is the one on the left. You need to look cross-eyed at this pair of images so that you effectively see three images. The middle one should appear three-dimensional. Some people can do this easily. Others need to trick their eyes into crossing.

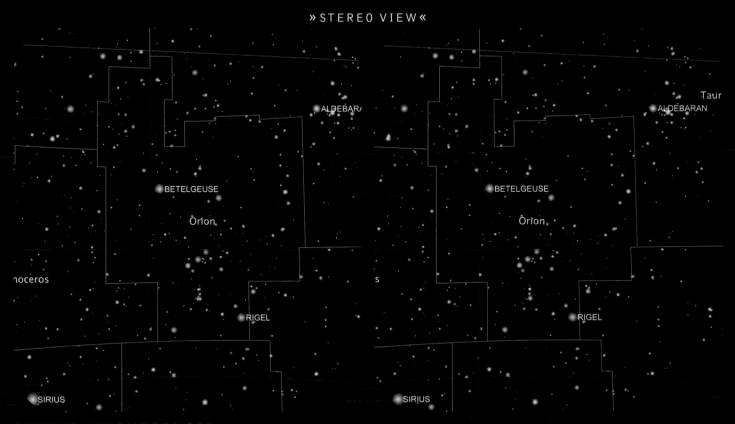

ORION'S NEIGHBORHOOD

Try the following method: Hold up one finger about 40 percent of the way between your eye and the page. Focus on the page. You will also see two blurry transparent images of your finger (one as seen by the right eye and one as seen by the left eye). Move your finger back and forth until these two transparent finger images are perfectly centered at the bottom of each image on the page. You might also need to tilt your head left or right to get the finger images to be level. Then, look directly at your finger. Behind your finger, you should see three versions of the picture on the page. If you carefully shift your gaze to the middle picture, without uncrossing your

eyes, it will magically appear as a beautiful 3-D image. Alternatively, start by holding up your finger as before, but now just look at your finger—your eyes will naturally cross to look at it. Behind it will appear three blurry versions of the picture on the page. Shift your gaze to the central one and it will appear in 3-D.

The previous page gives a 3-D view of the solar neighborhood seen from a distance. The sun can be seen as an ordinary, not so bright star, forming a right triangle in space with Sirius and Alpha Centauri; other stars and the faint Milky Way are in the background. This page shows a 3-D image of the constellation Orion.

To the Galaxies

ONCE WE HAD PARALLAX DISTANCES to stars, we could begin to construct a distance ladder to go from scale to scale. The stars are actually moving, but they are so far away that the motions are usually imperceptible. But if you wait long enough and compare before and after pictures, you can see the motions of individual stars in the sky. A star 100 parsecs away, traveling at 30 kilometers per second relative to the sun in a direction perpendicular to our line of sight (a tangential velocity), would move three seconds of arc in 50 years.

We can measure the velocity of stars along our line of sight (radial velocity) by taking their spectra. Stars show absorption lines in their spectra at specific wavelengths. Stars approaching us have their lines shifted to the blue— a blueshift; stars receding from us have their lines shifted to the red—a redshift. These Doppler effects are the same ones that cause a train whistle to go from high to low pitch as it first approaches us and then recedes into the distance. We can measure these Doppler shifts in the spectral lines of stars and determine their radial velocities with respect to us.

In a loose star cluster, if we extrapolate the proper motions of the stars into the future, we can see a convergence point in the sky where they are headed. The direction to this point tells us the ratio of the radial to the tangential velocities of the stars in the cluster. By measuring the radial velocities of the stars, we can figure out their tangential velocities. Comparing their tangential velocities with their observed proper motions, we can determine the distance to the star cluster. The brightness of main sequence stars in that cluster can then be used to set a distance scale. If we find similar main sequence stars appearing 100 times fainter in another cluster, we will know that it is 10 times farther away.

Harlow Shapley (1885–1972) found distances to nearby globular star clusters using RR Lyrae stars, whose brightness relative to main sequence stars could be calibrated in clusters. For more distant globular clusters, he used their sizes: A cluster half the angular size was twice as far away. He found the ensemble of globular clusters is centered on the center of our galaxy (about 8,000 parsecs, or 25,000 light-years, from us).

Milky Way

100,000
light-years across

FYI On average, Hubble figured, a galaxy half as big in angular size in the sky as another galaxy of the same type was roughly twice as far away.

In 1912 Henrietta Leavitt (1868–1921) found a relation between the brightness of Cepheid variable stars and their period of variability by observing a number of them in the Small Magellanic Cloud. They were all at approximately the same distance, so their relative luminosity as a function of their period of variability could be determined.

From 1923 to 1924 Edwin Hubble (1889–1953) observed the Andromeda galaxy (M31) with the 100-inch-diameter telescope on Mount Wilson and resolved M31 into stars. Hubble identified Cepheid variables in his sequence of photographs. But they were very faint, showing they were very far away. M31 is about 2.5 million light-years away. It is an entire galaxy like our own. When we look at the Andromeda galaxy, we are seeing it not as it is now but as it was 2.5 million years ago.

Hubble devised a number of methods for finding the distances to other galaxies: Cepheid variables, brightness of brightest stars, brightness of supernovae, and, for very distant galaxies, the angular size of the galaxy. Hubble then measured the radial velocities of these galaxies using the redshifts he found in their spectral lines. Galaxies showing a redshift are moving away from us. On average, Hubble found, the larger their distance, the larger their redshift. So galaxies are moving away from us. Moreover, the farther away they are, the faster they are moving. The universe is expanding!

By 1931 Hubble had found a distant galaxy receding at a speed of almost 20,000 kilometers per second. Today we know that, on average, for every million parsecs (a megaparsec) farther away a galaxy is from us, it will be going away from us an additional 71 kilometers per second faster. A galaxy that is 10 megaparsecs away from us will have a recessional velocity of 710 kilometers per second, and one that is 100 megaparsecs away from us will have a recessional velocity of 7,100 kilometers per second.

The galaxies are moving apart as the universe expands. Trace all those galaxies backward in time, and they all come together about 13.7 billion years ago—at the big bang.

The Milky Way and the Andromeda galaxy are shown at their true sizes relative to their distance apart.

2,500,000 light-years apart

Andromeda

110,000 light-years across

In cosmology, we are dealing with curved space-time, and there are several distance measures we might use. In most circumstances, we want to use a distance based on the look-back time. The universe is 13.7 billion years old. If we imagine galaxies having alarm clocks on them, each measuring the time elapsed since the big bang, a galaxy we see now with an alarm clock reading 12.7 billion years since the big bang would have a look-back time of 1 billion years. That's because light from it has been traveling through curved space for a billion years on its way to us. The distance derived from look-back time shows us where the galaxy was at the time we are seeing it—in the past. But by now it has moved farther away. How far away? We call that distance the co-moving distance.

We can measure a galaxy's look-back-time distance (how far away it is at the epoch when we see it) or its co-moving distance (how far away it has gotten to by now) by observing the redshift in its spectral lines and knowing the cosmological model. The redshift tells us how fast the galaxy is moving away from us because of the expansion of the universe. If we know the dynamic history of the universe, the galaxy's redshift can tell us its look-back-time distance or its co-moving distance today.

The Wilkinson Microwave Anisotropy Probe (WMAP) satellite has measured the cosmic microwave background in exquisite detail and combined these data with other data to produce an accurate cosmological model. If we take a snapshot of the universe at the present epoch, 13.7 billion years after the big bang, WMAP tells us that its geometry is approximately flat. That means we could make a good scale map of a slice of it extending out from Earth's Equator using a huge, flat sheet of paper. According to Einstein's equations of general relativity, the dynamic history of the universe depends on its energy content. The WMAP satellite data show the universe is currently composed of 4.6 percent normal matter; 0.01 percent thermal radiation left over from the big bang (cosmic microwave background radiation); 23 percent dark matter, made of unknown elementary particles, which cluster like the galaxies and provide the mass to bind clusters such as the Coma Cluster together; and 72 percent dark energy.

Dark energy is a quantum vacuum energy state, which gives the vacuum of empty space a constant energy density and a negative pressure. Since the negative pressure is uniform, it exerts no forces. In much the same way, the air pressure in the room where you are sitting is almost 15 pounds per square inch, but you don't notice it because it is nearly uniform.

The negative pressure produced by dark energy does have a gravitational effect, however, according to Einstein's theory of gravity—general relativity. Since it is a negative pressure (or suction), it exerts a negative gravitational effect, and since it operates in three directions (up-down, left-right, front-back), it has three times as large a gravitational impact as the positive gravitational effect of the energy supplied by the dark energy. Thus, dark energy produces an overall gravitational repulsion, causing the universe's expansion to be accelerating today.

Einstein first proposed this effect in 1917, calling it the cosmological constant. Adam Riess, Saul Perlmutter, Brian Schmidt, Bob Kirshner, Alex Filippenko, and their colleagues discovered it in 1997 when they carefully measured the expansion of the universe over time by measuring the distances to distant supernovae; they found that the universe's expansion is accelerating.

Vanderbei took a picture of M51 in May 2005. In June, a star in M51 exploded and became a supernova—so he took another picture.

M51
May 9, 2005

M51
July 10, 2005
Arrows point to supernova

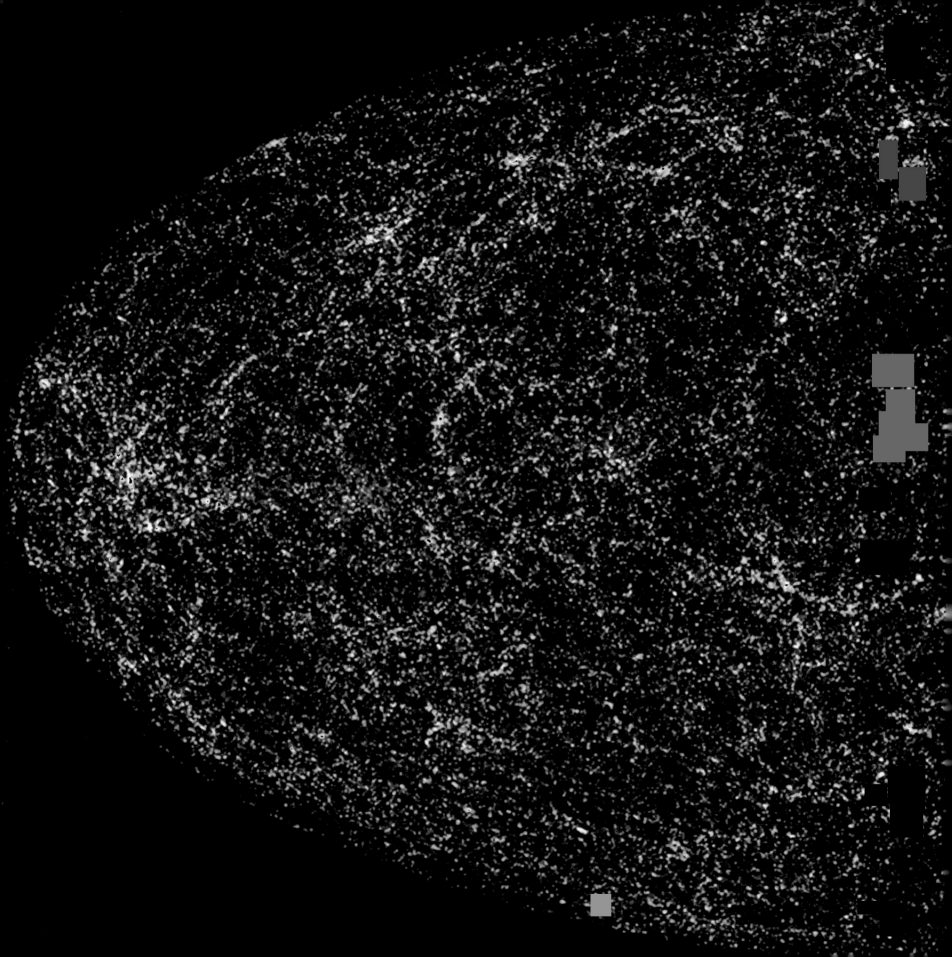

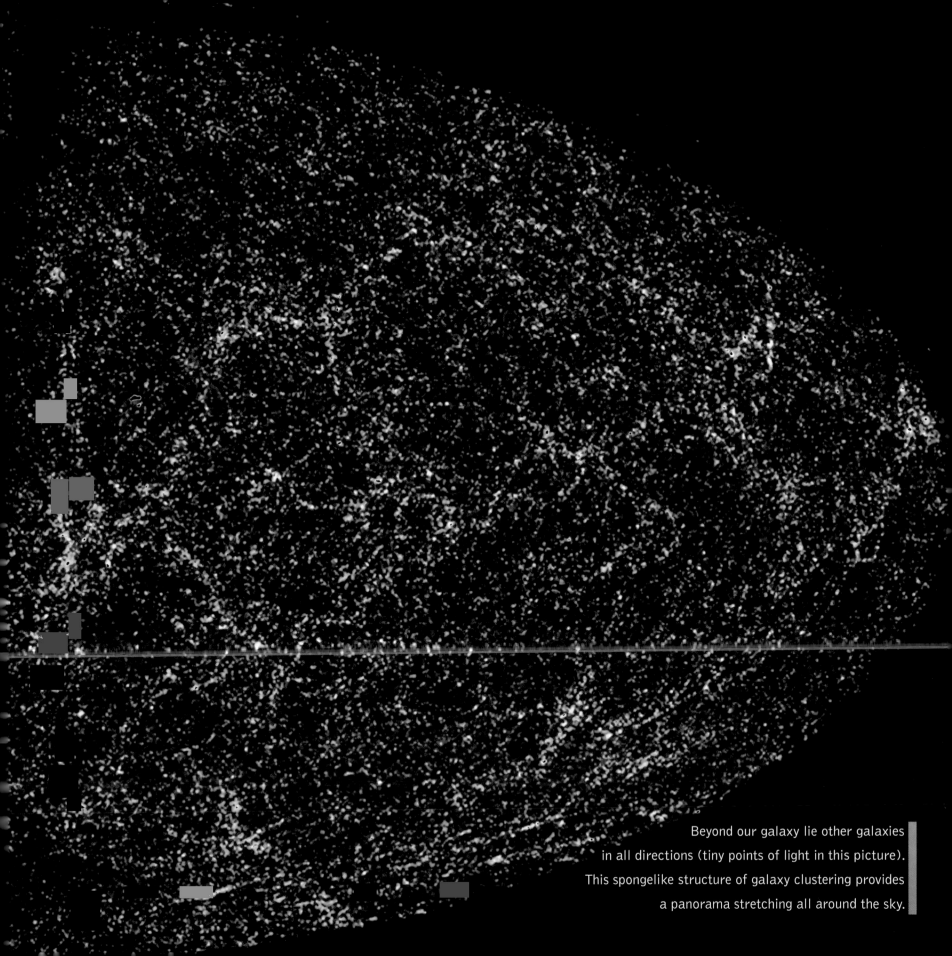

Beyond our galaxy lie other galaxies
in all directions (tiny points of light in this picture).
This spongelike structure of galaxy clustering provides
a panorama stretching all around the sky.

To the Edge of the Visible Universe

ARMED WITH OUR COSMOLOGICAL model from WMAP satellite data, we can calculate the look-back-time distances of objects, whose redshifts we can measure, out to the edge of the visible universe. The Subaru Deep Field galaxy is the most distant galaxy known—12.8 billion light-years away. The current record holder is gamma-ray burster GRB 090423, discovered by the Swift satellite at a distance of 13.07 billion light-years.

The most distant thing we can see is the cosmic microwave background. It was emitted when the universe was about 1,090 times smaller than it is today, when the universe was only 380,000 years old. These cosmic microwave background photons have come to us directly from an epoch almost 13.7 billion years ago. Since then, they have traveled almost 13.7 billion light-years through expanding space to reach us. So the look-back-time distance to the cosmic microwave background is 13.7 billion light-years.

These cosmic microwave background photons scattered off protons and electrons just before heading on their long journey to Earth. Those protons and electrons have been moving outward ever since and by now would be 45.6 billion light-years away from us. That co-moving distance is larger than 13.7 billion light-years because the space between us and those protons and electrons has been stretching so fast that the distance separating them from us has been lengthening at a speed faster than the speed of light. No objects are passing each other locally at speeds faster than the speed of light, upholding special relativity. It's just that the space between two distant objects is stretching faster than the speed of light—something allowed by Einstein's theory of general relativity. Thus, we can plot objects in the universe out to a current maximum distance of 14.0 billion parsecs.

Now that we know the distances to different objects in the universe, ranging from planets to stars to distant galaxies, we are ready to make a map of the universe.

On the following two pages, we show the WMAP map of the cosmic microwave background radiation, a faint glow of microwaves that come to us from all over the celestial sphere—from every direction. We are effectively living inside a giant, expanding microwave oven, but it's a rather low temperature oven now—only 2.7 degrees above zero on the Kelvin scale (that's minus 270°C). This microwave background radiation is uniform in temperature to about 1 part in 100,000 all over the celestial sphere. This is very smooth—if the radius of the Earth varied by only 1 part in 100,000 over the globe, mountains would be only 61 meters (200 feet) high!

But small as they are, these fluctuations, visible to the WMAP satellite, show the fingerprint of creation—a baby picture of the universe as it appeared just 380,000 years after the big bang. The hot spots in the microwave background are shown in red and the cold spots are shown in blue, with intermediate temperatures in green. These slight hot and cold spots—fossil remnants of quantum fluctuations produced much earlier, when the universe was less than 10^{-34} seconds old—represent density variations that eventually grew by gravity to form the great clusters of galaxies we see today. The pattern of fluctuations supports Alan Guth's 1981 theory of inflation, which says the early universe was dominated by a great amount of dark energy whose repulsive gravitational effects started an explosive expansion—the big bang.

Subaru Deep Field galaxy, the most distant known galaxy in the universe, is a tiny red dot at the center of the inset picture.

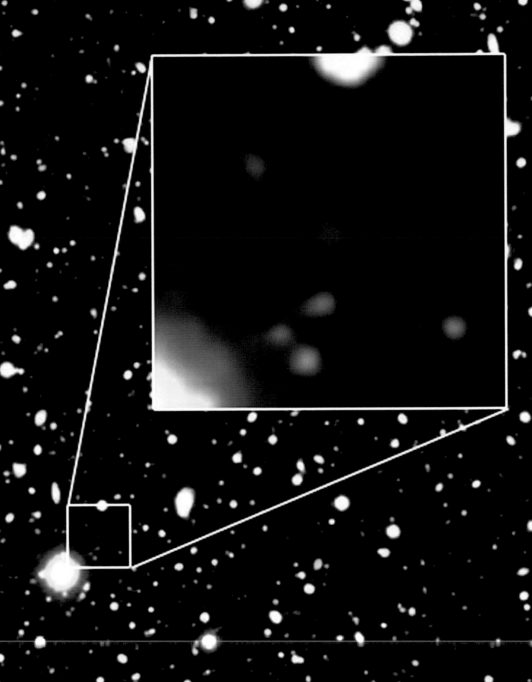

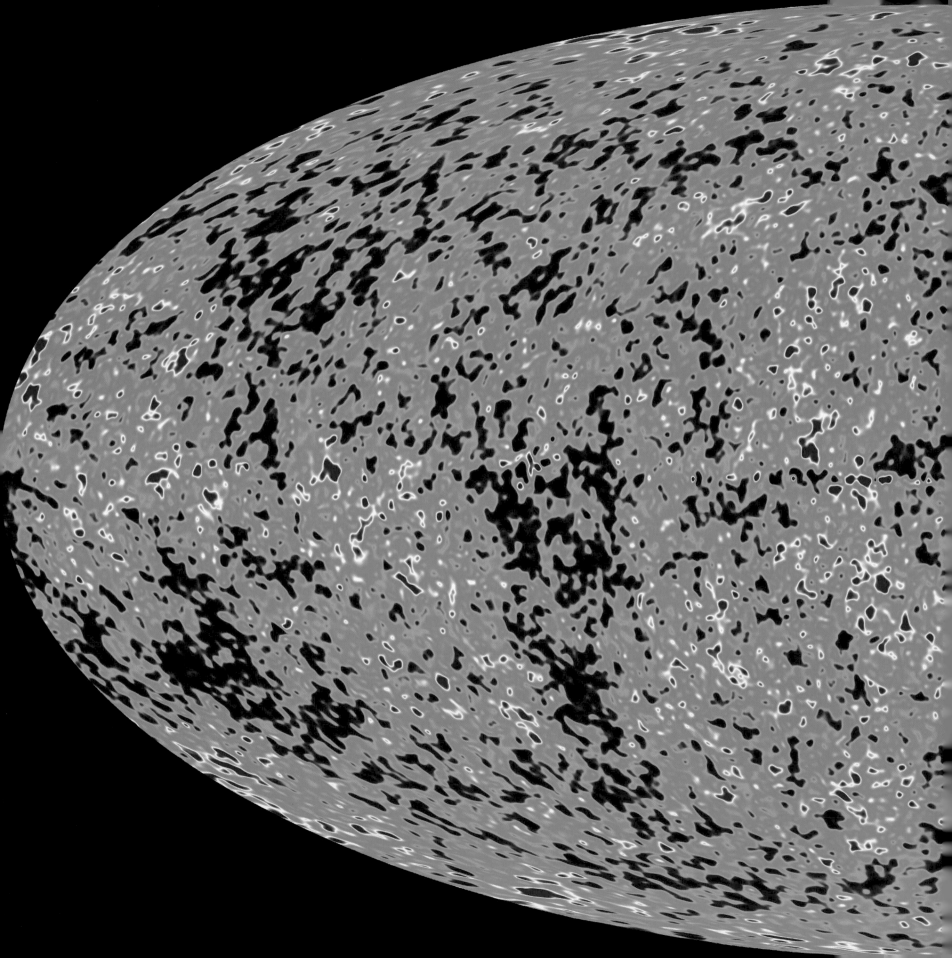

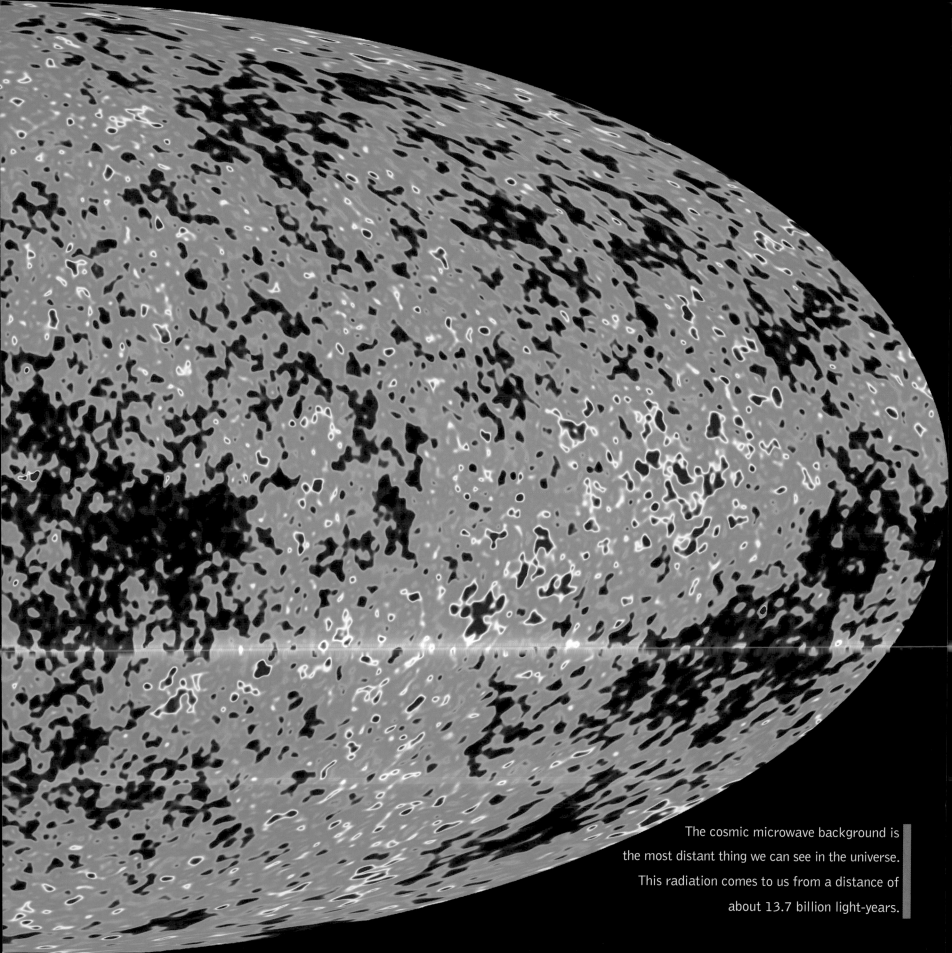

The cosmic microwave background is
the most distant thing we can see in the universe.
This radiation comes to us from a distance of
about 13.7 billion light-years.

MAP OF THE

CHAPTER 4

UNIVERSE

A COSMIC SURVEY

IN CHAPTER 2, we mapped the celestial sphere. Can we go beyond these maps of the sky? Can we create maps that show even more of what we know about the universe? We would like to survey the universe in 3-D (in depth) to show how far away things are. But how can we show the universe in depth on a flat piece of paper?

In 1985 Valérie de Lapparent, Margaret Geller, and John Huchra pioneered making slice maps of the universe. They surveyed a thin slice of sky, 117 degrees long and 6 degrees wide and, using redshifts, measured the distances to all the galaxies in the slice. They could then make a fan-shaped diagram of what they saw.

Earth is positioned at the vertex of the fan, and each galaxy is placed at the correct relative distance from Earth, with its angular position in the sky properly illustrated along the left-to-right angular arc of the fan. Plotting galaxies on a fan-shaped map enabled Geller and Huchra to discover the CfA Great Wall of galaxies (named after their Harvard-Smithsonian Center for Astrophysics survey). This Great Wall of galaxies is 758 million light-years long and was recognized by the *Guinness Book of Records* as the largest structure in the universe discovered up until then. When the maps came out, they created quite a sensation, as no one had expected to find structures this large.

Gott, Mario Jurić, and our colleagues made an equatorial-slice map of the 126,594 galaxies and quasars within two degrees of the celestial equator in the Sloan Digital Sky Survey. In this equatorial slice was an even larger wall of galaxies, which we named the Sloan Great Wall. We measured its length to be 1.37 billion light-years. As mentioned in the Preface, it was included in the *Guinness World Records 2006* as the largest structure in the universe, breaking the previous record set by the CfA Great Wall.

The Sloan Great Wall and the CfA Great Wall are structures that began as small random variations in the amount of energy in the early universe, about 10^{-37} seconds after the big bang. At that time, the part of the universe visible at a given location was tiny—only about 10^{-37} light-seconds (or 3×10^{-27} cm) in radius—because that is as far as light (traveling at 3×10^{10} cm per sec) can travel in the 10^{-37} seconds since the beginning. On such small scales, random quantum variations in the amount of energy seen from place to place occurred due to the uncertainty principle of quantum mechanics.

These regions of above- or below-average energy have stretched greatly as the universe has expanded. When we look at the cosmic microwave background, we can see these energy variations (of order 1 part in 100,000) as they appear at an epoch 380,000 years after the big bang. In the subsequent 13.7 billion years, regions with greater than average density will be pulled together by gravity and expand less rapidly than regions of less than average density. These variations in density will grow in amplitude by the action of gravity as the denser regions accumulate more matter and the less dense regions thin out. The above average density regions become the great walls and great clusters of galaxies we see today.

These most dramatic structures in the universe are thus the fossil remnants of tiny random quantum fluctuations in energy produced in the first moments of the

Sloan Great Wall of galaxies (top fan) and CfA Great Wall of galaxies (bottom fan), shown at the same scale. Individual galaxies are shown as dots. Earth is at the vertex at the bottom of the bottom fan. Right ascension is celestial longitude. Recessional velocities are proportional to distance from us.

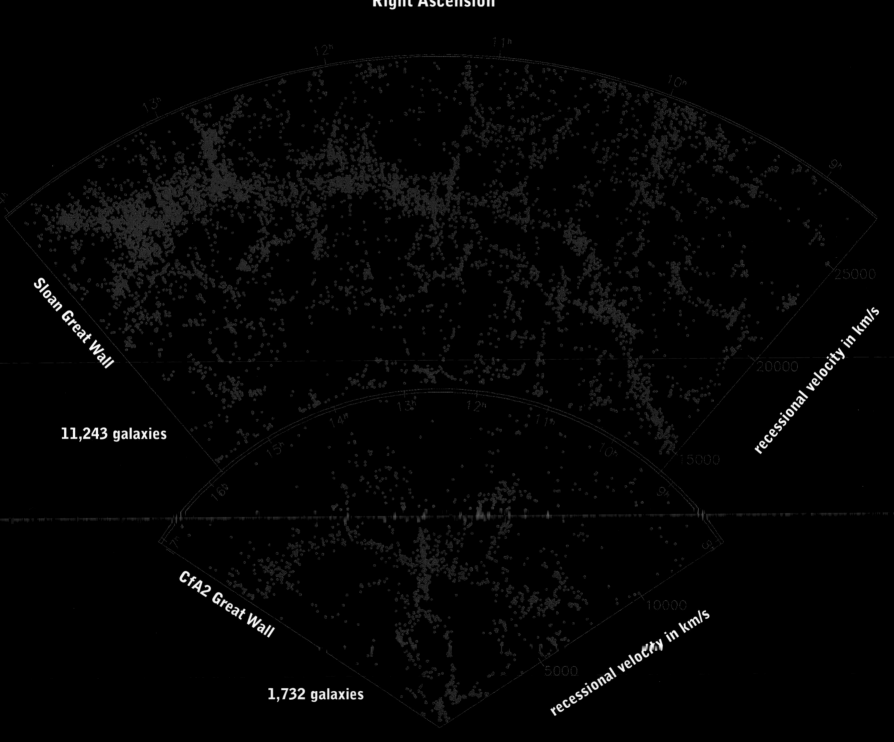

Right Ascension

Sloan Great Wall

11,243 galaxies

recessional velocity in km/s

25000

20000

15000

CfA2 Great Wall

1,732 galaxies

recessional velocity in km/s

10000

5000

universe's history. Since the original quantum fluctuations are random, the high- and low-density regions must effectively be interchangeable. This produces a spongelike geometry (as first argued by Gott, Adrian Melott, and Mark Dickinson in 1986).

A marine sponge has many water passageways that percolate through it, bringing nutrition to all its parts. If we were to pour concrete into the water and let it harden, it would produce a concrete sponge—complementary to the original one. As gravity enhances the contrast between the high- and low-density regions, this interlocking spongelike geometry of high- and low-density regions remains. High-density filaments like the Sloan Great Wall connect the great clusters to each other, and giant, empty voids are likewise connected to other voids by low-density tunnels.

As the universe continues to expand, the Sloan Great Wall will retain its general shape but continue to expand in size. Gravity will continue to enhance its density as it pulls nearby galaxies toward it. The Sloan Great Wall is like the crest of a giant wave where galaxies are brought crashing together by the action of gravity.

But its two ends will continue to move apart as they continue to participate in the general expansion of the universe. As this expansion accelerates in the future, observers at opposite ends of the wall would eventually lose sight of each other as the light passing from one to another is unable to cover the ever more rapidly stretching space between them. Einstein's theory of special relativity says you can't build a rocket ship traveling faster than a beam of light, but the space between two galaxies can stretch so fast that a light beam cannot complete the journey between them.

Each galaxy in the Sloan Great Wall harbors approximately a hundred billion stars. More than 10,000 galaxies reside in the fan-shaped region occupied by the Sloan Great Wall. With galaxies shown the size you see opposite, it would take a map 76 meters (250 feet) across to show the complete Sloan Great Wall. To be shown at proper scale, the distances between the galaxies in the picture would have to be stretched by a factor of 50. To show the entire visible universe on this scale would require a map more than 1.13 kilometers (3,707.35 feet) across. Such are the vast sizes we encounter in the universe. And yet on this enormous map, our entire solar system out to Neptune would be small enough to fit inside an atom. Making a map of the universe is quite a challenge.

Perhaps the first book to address this challenge was *Cosmic View: The Universe in 40 Jumps* by Kees Boeke, published in 1957. This brilliant book started with a picture of a little girl sitting in her school courtyard shown at one-tenth scale. The next picture showed the same little girl, at one-hundredth scale. Each successive picture appeared at ten times smaller scale, and therefore showed a ten times wider view. The sequence continued until the 26th and last picture, which showed galaxies out to a distance of 750 million light-years—the greatest distances surveyed at that time. (We've now seen out 18 times farther.) A further sequence of pictures showed microscopic views, each with ten times greater magnification. But Boeke's book is still an atlas of maps, not a single map. How does one show the entire observable universe in a single map?

Starting at right and on the following pages is a panorama of the eastern end of the Sloan Great Wall. Individual galaxy images have each been enlarged by a factor of 50 so that they can be seen easily.

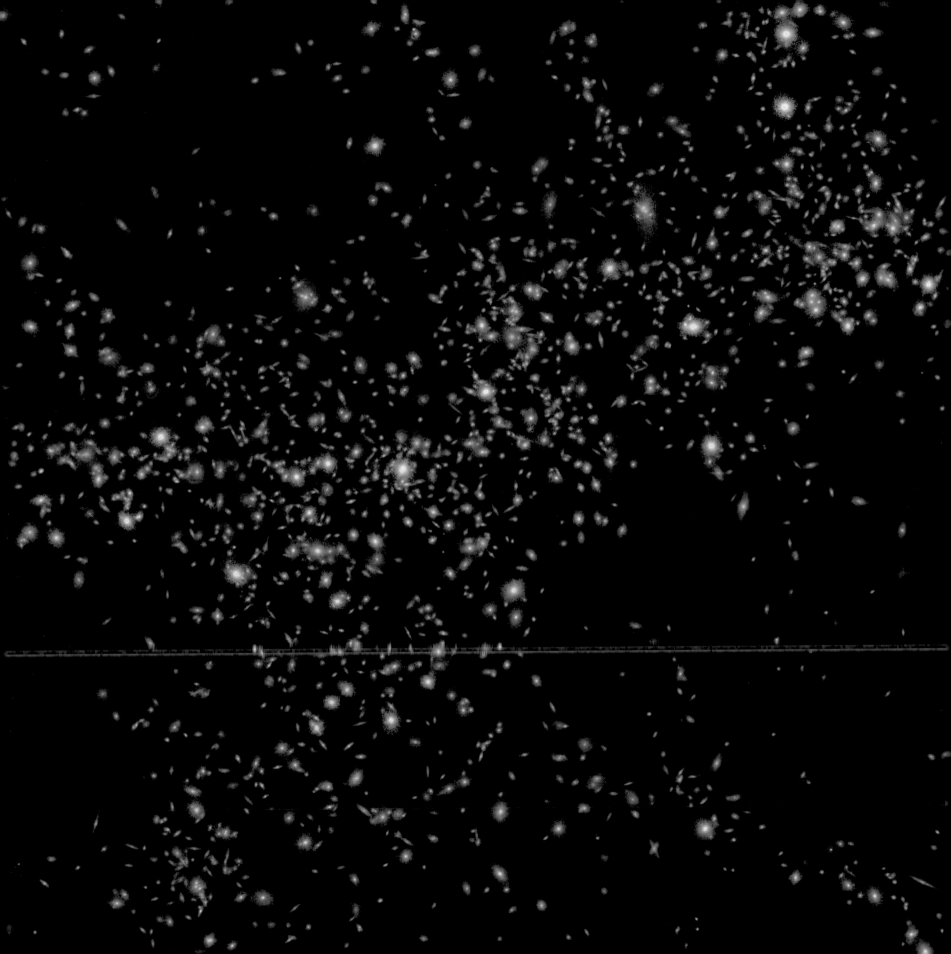

Here and continuing from the preceding page is a continuous panorama by Lorne Hofstetter and J. R. Gott of the eastern end of the Sloan Great Wall. Galaxy images are enlarged 50× relative to the distances between them. To show these galaxies (as pictured) at their true separations, the panorama would have to be 125 feet long.

ALL IN ONE

AS WE HAVE SEEN, astronomers mapping the universe are confronted with the challenge of showing a wide variety of scales. What should a map of the universe include? It should show locations of all the famous things in space: the Hubble Space Telescope, satellites orbiting Earth, the moon, the sun, planets, asteroids, Kuiper belt objects, stars such as Alpha Centauri, black holes, galaxies, quasars, and finally the cosmic microwave background radiation itself.

We want our Map of the Universe to display clusters of galaxies, but also stars within our own galaxy and the sun, moon, and planets. Objects close to us may be inconsequential in terms of the whole universe, but they are important to us. This calls to mind the famous *New Yorker* cover of March 29, 1976, "View of the World from 9th Avenue" by Saul Steinberg, which humorously depicts a New Yorker's view of the world. The traffic, sidewalks, and buildings along Ninth Avenue are visible in the foreground. Behind is the Hudson River, with New Jersey as only a thin strip on the far bank. At even smaller scale, we see the rest of the United States, with the Rocky Mountains sticking up like small hills. In the background, but not much wider than the Hudson River, is the entire Pacific Ocean, with China and Japan in the distance.

This is, of course, a parochial view, but it is just the kind of view that we are looking for. For our Map of the Universe, we would like a single map that would show equally well both interesting objects in the solar system, nearby stars, galaxies in the local group, and large-scale galaxy clustering out to the cosmic microwave background.

Armed with the standard cosmological model from the Wilkinson Microwave Anisotropy Probe (WMAP) data, we can plot the expansion of the universe as a function of time. For the first time, we have fairly accurate knowledge of the geometry of the universe and the distances to individual galaxies and quasars, so we can plot a map of where they will be on a certain date.

In 1972, while I (Gott) was a graduate student, I invented a map projection for the universe and produced small versions of it over the years. Confronted with the task of displaying the galaxy-clustering data from the Sloan Digital Sky Survey, Mario Jurić and I decided that it was time to make a large-scale detailed version of my

Map of the Universe to summarize all the exciting new discoveries that had been made recently in astronomy. The map projection I had devised was conformal, preserving local shapes, as the Mercator projection does, while still allowing us to cover the wide range of scales—from Earth's neighborhood to the cosmic microwave background. We include all objects known within two degrees of the celestial equator—plus famous objects above and below the celestial equator.

FYI This map projection is conformal because on the map of the pizza slice (far right) the pepperoni retain their true round shapes.

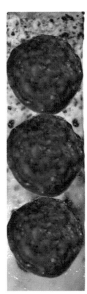

The WMAP satellite data tell us that the geometry of such a two-dimensional cross section taken through the universe at the present epoch (using co-moving distances) is flat like a piece of paper. The map of the universe shows a 360-degree panorama from left to right, looking out from Earth's Equator, while the vertical coordinate shows distance from Earth. Earth's surface is a horizontal line at the bottom of the map, and the cosmic microwave background is at the top. As you move up the map, you get farther and farther from Earth. The left- and right-hand borders coincide; they could be taped together to display the map as a cylinder. The horizontal coordinate is celestial longitude (which divides the 360-degree view into 24 hours—each hour covering 15 degrees).

Lines fanning out from Earth's Equator, like edges of a pizza slice, are shown as parallel vertical lines in the map. Imagine a round piece of pepperoni sitting on a slice of pizza just wide enough to span the width of the slice (above left). The slice gets wider the farther it gets from the pie's center, so if there was another piece of pepperoni farther

out that also spanned the pizza slice, it would have to be bigger. Those two pieces of pepperoni would be shown with the same width on the map because they cover the same angle as seen from the center of the pizza. In our pizza slice we show three pieces of pepperoni. The first one, nearest the center, is the smallest. The second one is twice as big as the first, and twice as far away. The third is twice as big as the second and twice as far away. Now, usually a pizza will have pieces of pepperoni that are all the same size, but in our special pizza, the pepperoni pieces get bigger farther out. The strip on the right shows our map projection of the same pizza slice. Radial lines, the edges of the wedge-shaped pizza slice, show up as vertical lines on the map.

If the map is to be conformal, showing local shapes well, then the pepperoni pieces should all be shown as

Saul Steinberg's famous *New Yorker* cover (opposite) provides an illustration of scale, as does a slice of pepperoni pizza (above left). The same slice after applying a logarithmic conformal map (above right).

approximately round, since their true shapes are round. On the map they would appear as three circles of equal size, one just above the other. The vertical coordinate shows distance from the center. If one had a sequence of such pepperoni pieces—each larger than the last by a constant factor, each exactly spanning the ever widening pizza slice, they would show up as a stack of equal-size circles on the map, and the pizza slice itself would show up as a rectangular vertical strip. Each piece of pepperoni spans the same angle as seen from the center. So in the Map of the Universe, objects that span the same angle in the sky are shown at the same size.

Equal factors of distance farther away in the pizza slice (a factor of two, in the case of our pepperoni) are shown as constant vertical increments in the map (the diameter of one pepperoni in the map). This is called a logarithmic scale. Each factor of ten farther out takes us a constant additional vertical distance upward on the map. Objects that are ten times farther away are shown at one-tenth the scale. This allows the Map of the Universe to plot distant objects such as galaxies and quasars on the same map as solar system objects.

We are familiar with logarithmic scales in everyday life. For example, the Richter scale for earthquakes is a logarithmic scale. An earthquake of magnitude 8.0 on the Richter scale is ten times more energetic than one of magnitude 7.0, which in turn is ten times more energetic than one of magnitude 6.0, and so forth.

In sum, the Gott-Jurić map is easy to describe: The horizontal scale from left to right shows angles in equal increments in celestial longitude, and the vertical scale is logarithmic in distance from the center of Earth. It is a conformal map, showing local shapes perfectly. Objects of the same angular size are shown at the same size on the map, and objects ten times farther away are shown at one-tenth

the size. Thus, the sun and moon are shown at equal size on this map. They are each one-half degree across on the sky and should each take up 1/720th of the width of the map, which spans the full 360-degree panorama. The sun is 400 times farther away than the moon, and so by the rules of the map, it appears at 1/400th the scale. However, because the sun is actually 400 times larger than the moon, it appears the same size as the moon on our map.

Having objects of the same angular size appear at the same size on the map makes psychological sense. If you stand on Ninth Avenue in New York, the buildings in front of you subtend an angle just as large to your eye as would distant California. So the New Yorker's view of the world is plausible psychologically. Things that are closer to us are more important to us. Similarly, because the size of an object on the map is proportional to its angular size in the sky, objects that are more prominent in the sky would be more prominent in the map.

In this book, we are presenting a Map of the Universe that is almost 40 inches tall, displayed on a four-page gatefold. The sun, moon, and other famous objects would be too small to be seen easily on these maps (because the universe is mostly empty space), and so we supply blown-up pictures of each. The map shows a snapshot of the universe at 4:48 a.m. Greenwich Mean Time (GMT), August 12, 2003. It shows planets at that time, and for distant galaxies it shows them at the distances they will have attained from Earth by that time (their co-moving distances).

Discovered by William Herschel (who also discovered Uranus), the Eskimo Nebula, seen in this beautiful Hubble Space Telescope image, is a dying solar-type star shedding gas.

MAP OF THE UNIVERSE

Inside this gatefold is the Gott-Jurić Map of the Universe. Open it and rotate the book 90° counterclockwise.

FROM SIDE TO SIDE, the map represents a 360° panorama looking out from Earth's Equator. The surface of Earth at the Equator is shown as the straight line near the bottom of the map. The vertical coordinate shows distance from the center of Earth. Moving upward, each large tick mark on the right shows a distance that is ten times farther away. Regions that are ten times farther away are shown at one-tenth the size so that the map can include everything from the surface of Earth to the cosmic microwave background while still preserving shapes locally. In other words, each cloud of asteroids or wall of galaxies is shown in its proper shape—neither stretched

nor squashed vertically. The map shows objects' positions as of 4:48 a.m. GMT, August 12, 2003. We plot all objects within two degrees of the equatorial plane plus famous objects outside that plane.

Just above the surface of Earth you can see all 8,420 artificial satellites in orbit. Since radial lines spreading out from the center of Earth separated by an angle (of, say, one-half degree) are shown as parallel vertical lines in the map, objects should be shown at a size on this map that is proportional to their angular size. But they would be tiny. The sun and moon, for example, each one-half degree across, would cover only 1/720th of the diameter of the map,

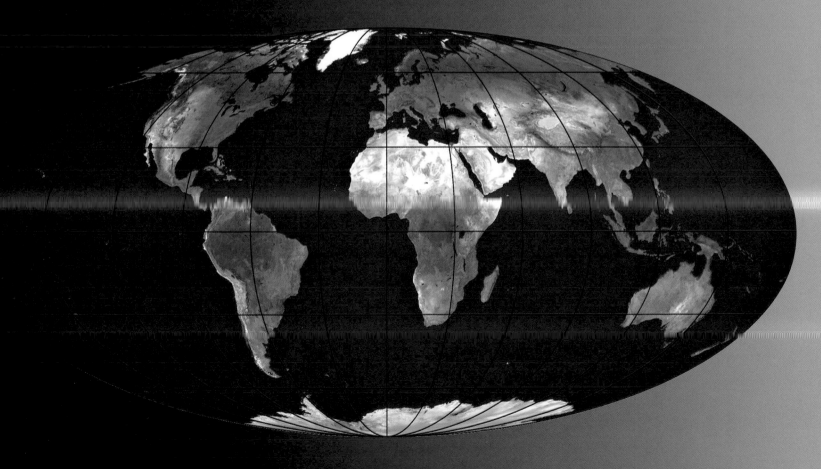

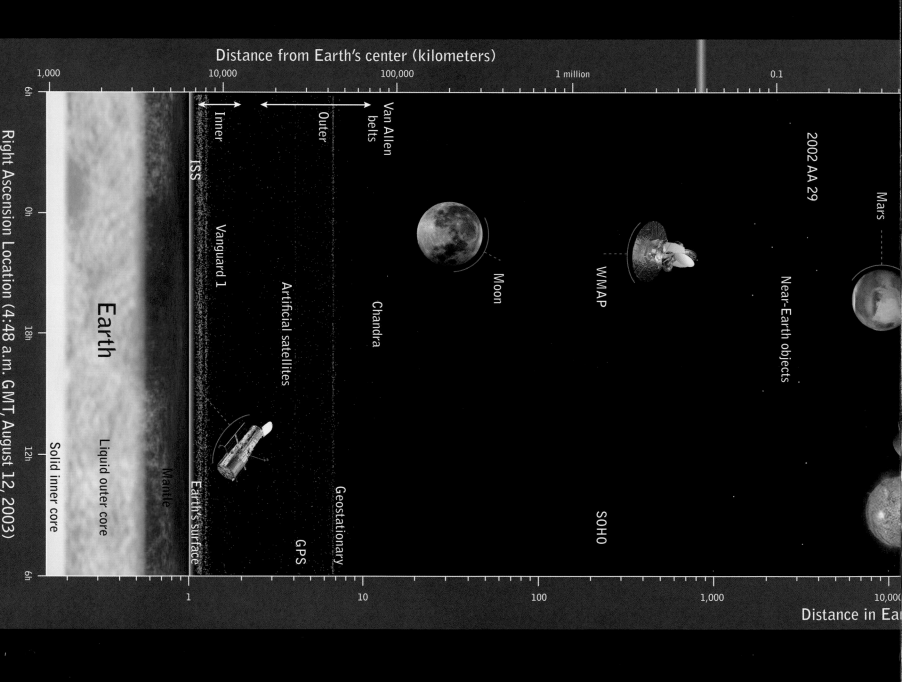

Distance from Earth's center (kilometers)

1,000 | 10,000 | 100,000 | 1 million | 0.1

Right Ascension Location (4:48 a.m. GMT, August 12, 2003)

6h | 0h | 18h | 12h | 6h

Solid inner core

Liquid outer core

Mantle

Earth

Earth's surface

ISS

Inner

Outer

Van Allen belts

Vanguard 1

Artificial satellites

Chandra

Moon

GPS

Geostationary

WMAP

SOHO

2002 AA 29

Near-Earth objects

Mars

1 | 10 | 100 | 1,000 | 10,000

Distance in Ear

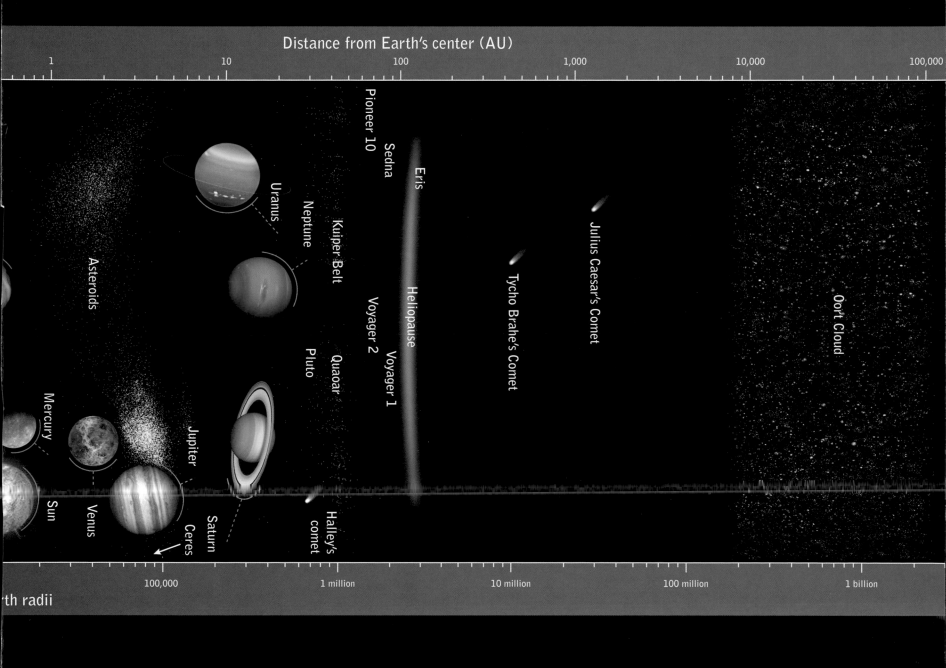

Distance from Earth's center (AU)

1   10   100   1,000   10,000   100,000

Pioneer 10

Sedna

Eris

Uranus

Neptune

Kuiper Belt

Julius Caesar's Comet

Asteroids

Heliopause

Tycho Brahe's Comet

Oort Cloud

Voyager 2

Voyager 1

Pluto

Quaoar

Mercury

Jupiter

Saturn

Venus

Ceres

Halley's comet

Sun

100,000   1 million   10 million   100 million   1 billion

rth radii

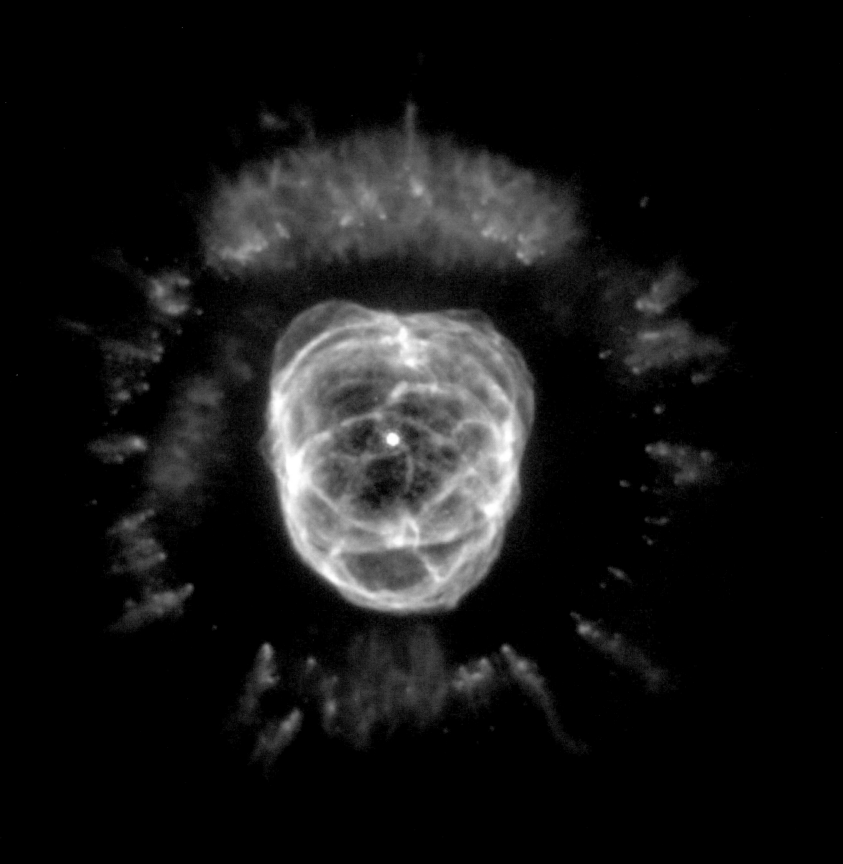

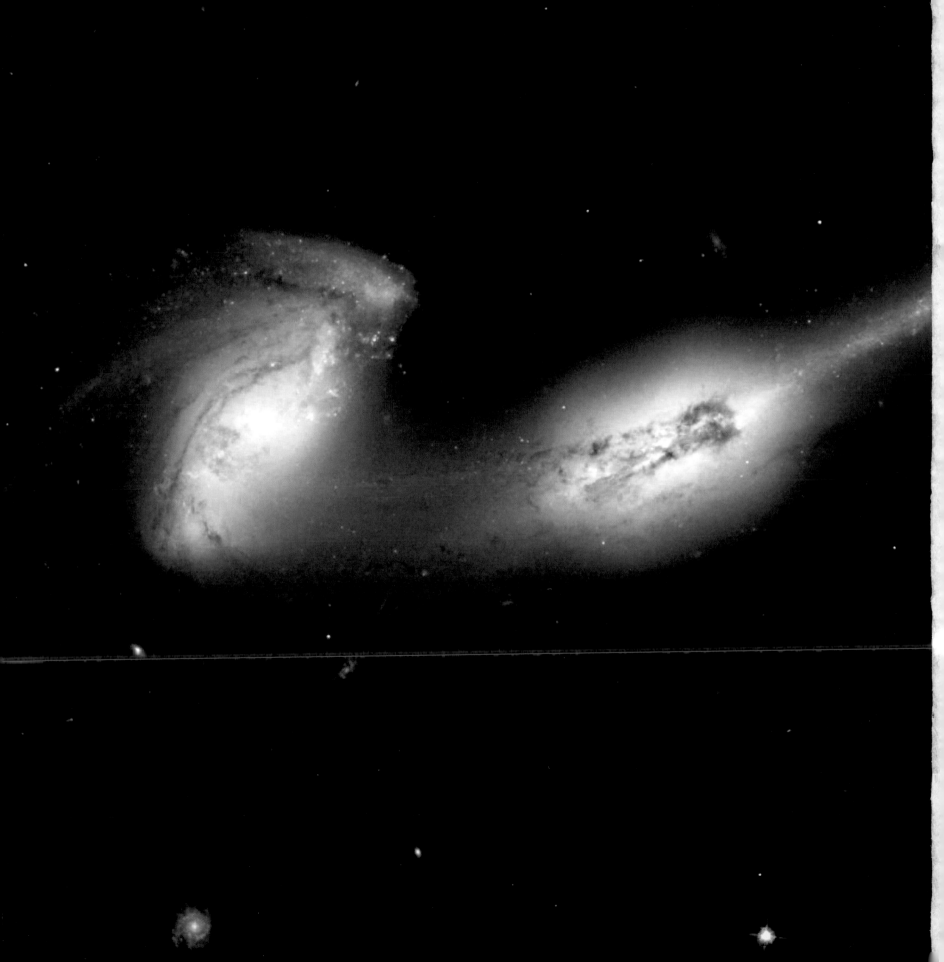

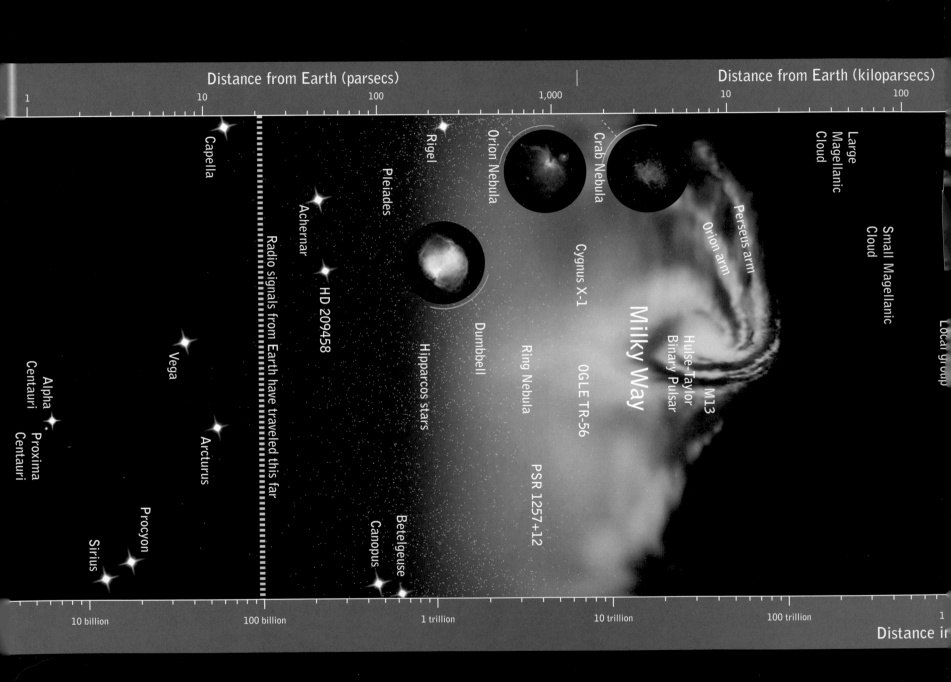

1   10   100   1,000   10   100

Capella

Rigel

Orion Nebula

Crab Nebula

Large
Magellanic
Cloud

Pleiades

Achernar

Radio signals from Earth have traveled this far

HD 209458

Perseus arm

Orion arm

Small Magellanic
Cloud

Local group

Cygnus X-1

Milky Way

Vega

Hipparcos stars

Dumbbell

Ring Nebula

OGLE TR-56

Hulse-Taylor
Binary Pulsar

M13

Alpha
Centauri
Proxima
Centauri

Arcturus

Procyon

Sirius

Canopus

Betelgeuse

PSR 1257+12

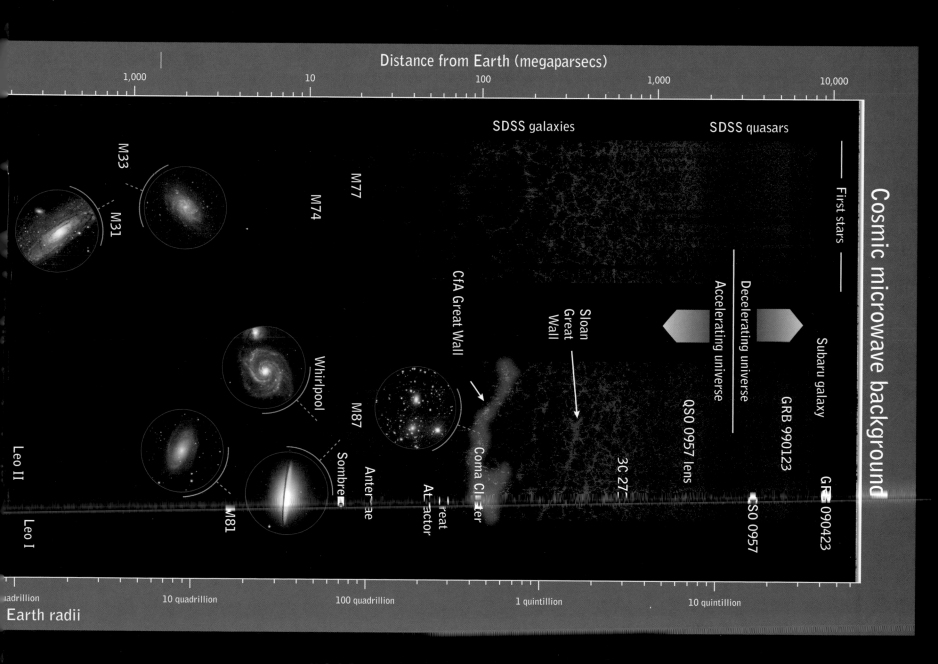

Distance from Earth (megaparsecs)

1,000 10 100 1,000 10,000

Cosmic microwave background

SDSS galaxies   SDSS quasars

First stars

M33
M31

M77
M74

CfA Great Wall

Sloan Great Wall

Decelerating universe
Accelerating universe

Subaru galaxy

GRB 990123

Whirlpool

M87

Sombrero

Antennae

Coma Cluster

Great Attractor

3C 27

QSO 0957 lens

QSO 0957

GRB 090423

Leo II

M81

Leo I

Earth radii

adrillion 10 quadrillion 100 quadrillion 1 quintillion 10 quintillion

Note: Inset images are magnified for clarity.

which covers 360°. Therefore, we have included blown-up pictures of famous objects. The moon, at 60 Earth radii, is as far as humans have gone. Mars is making a close approach to Earth—9,000 Earth radii away. Mercury, the sun, and Venus are next as we move up the map and farther away from Earth. The asteroid belt appears as a wavy band, with clumps where it crosses this slice extending from Earth's Equator. We pass Jupiter, Saturn, Uranus, Neptune, and the Kuiper belt as we leave the solar system.

The Oort cloud, where comets originate, is next, followed by nearby stars such as Alpha Centauri and Sirius. Farther away, one can see the spiral arms of our own Milky Way galaxy, and beyond that are nearby galaxies such as M31 and M81. The CfA Great Wall of galaxies (0.758 billion light-years long) appears, and above it we see the Sloan Great Wall (1.37 billion light-years long)—about twice its size. But the Sloan Great Wall appears smaller here because it is 2.5 times farther away and is therefore displayed at 2.5 times smaller scale. The 126,594 Sloan Survey galaxies and quasars are depicted in two vertical bands. The spaces in between are zones of avoidance where the plane of our own galaxy crosses the plane of Earth's Equator, blocking the view of distant galaxies. These zones were not included in the Sloan Survey. At the top of the map is the cosmic microwave background, the most distant thing we can see.

Versions of our map have been reprinted in the *New York Times, New Scientist,* and *Astronomy* magazine— more than 1.5 million copies in all. The map shown in the gatefold is a new full-color version almost 40 inches long.

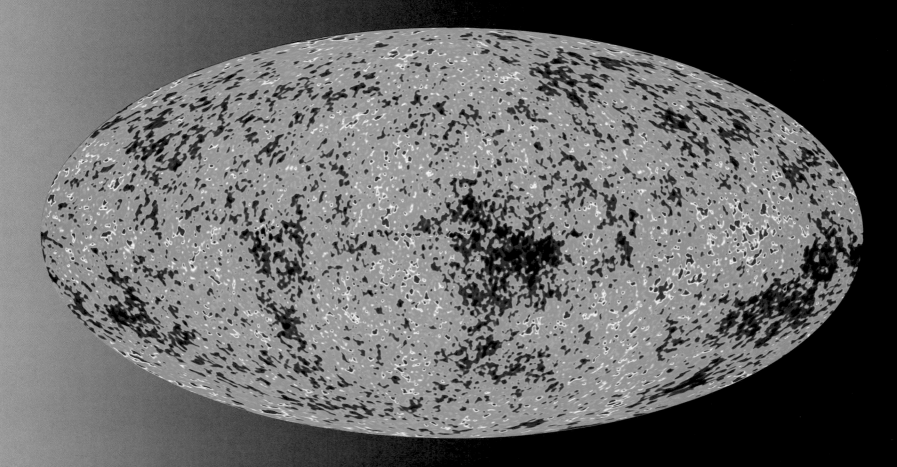

Landmarks

THE LINE SHOWING the surface of Earth is depicted as perfectly straight, because on this scale the altitude variation in Earth's surface is too small to be visible.

Earth's atmosphere is shown in shades of blue above Earth's surface. The ionosphere (blue) occupies an altitude range of 70 to 600 kilometers (40 to 370 miles) above Earth's surface. The ionosphere marks the practical outer extent of Earth's atmosphere. Auroras are a prominent visible feature of the ionosphere. Below the ionosphere is the stratosphere, occupying an altitude range of 12 to 50 kilometers (7 to 30 miles). The troposphere (0 to 12 kilometers, 0 to 7 miles), where weather occurs, is so narrow that it is subsumed into Earth's surface-line thickness. This map shows properly just how narrow the atmosphere is relative to the circumference of Earth.

We have shown all 8,420 artificial Earth satellites (not just those in the equatorial slice) in orbit as of August 12, 2003, 4:48 a.m. GMT. This is the last full moon before the closest approach of Mars to Earth in 2003. The time was chosen for the placement of the sun, moon, and Mars.

These are actual named satellites, not just space junk. The great majority of Earth satellites are in low Earth orbit, skimming just above Earth's atmosphere. The International Space Station and the Hubble Space Telescope are both in low Earth orbit. Because of atmospheric drag, the International Space Station needs occasional missions to boost it in altitude and keep it in orbit. Some of the lowest satellites are on their way toward reentry now. Vanguard 1, launched in 1958, is the earliest launched satellite still in orbit.

Global positioning system satellites (which provide position information to GPS navigation devices) cover the globe in nearly circular orbits at identical altitudes, making a line on the map.

We also see a line of geostationary satellites at an altitude of 35,000 kilometers (22,000 miles) above Earth's surface. At this altitude, the orbital period of a satellite exactly matches the rotation period of Earth: These satellites are placed in equatorial orbit going east, so they will stay positioned over a fixed spot on Earth's equator. The inner and outer Van Allen radiation belts are bands of charged particles trapped by Earth's magnetic field. The Chandra X-ray Observatory, which studies black holes and clusters of galaxies, is also shown.

At approximately 60 Earth radii, the moon marks the extent of direct human occupation of the universe. Behind the full moon at approximately four times its distance from Earth is the WMAP spacecraft. On the opposite side of Earth, 180 degrees away (to the right) on this panorama, is the SOHO satellite, which continuously observes the sun.

The next objects are the closest asteroids (called near-Earth objects). Asteroid 2002 AA29 is about 0.1 kilometer (0.06 mile) in diameter and follows a nearly circular orbit at 1 AU from the sun, just like Earth. It makes close approaches to Earth every 95 years but is in no danger of colliding in the near future.

However, in general, near-Earth objects do pose a danger to us on Earth. An asteroid about ten kilometers (six miles) across hit Earth 65 million years ago, killing off the dinosaurs. A strike by a 1-kilometer-diameter (0.6-mile) asteroid could cause a civilization-ending event—expected to occur about once every million years. We are currently on the lookout for such objects. With enough

---

The Mice, a pair of colliding galaxies about 300 million light-years away

warning (say 10 to 100 years before a projected collision), a space mission might give the approaching asteroid a small kick to divert it from hitting Earth.

The next object is Mars, at a distance of approximately 9,000 Earth radii. In general, we will show all objects of a given class (such as asteroids) within two degrees of the celestial equator, plus famous individual objects (such as Mars), even if they are above or below the equatorial slice. Mars is shown near its point of closest approach (which it achieved on August 27, 2003, at a center-to-center distance of 55,758,005 kilometers). Farther from us are Mercury, the sun, and Venus. The sun is 180 degrees away from the moon, or halfway across the map horizontally, since the moon is full.

Snaking across the map is the asteroid belt, where most asteroids are found. Of the total of 218,484 asteroids in the ASTORB database, we have shown only those 14,183 that are within two degrees of the celestial equator—that

is, directly above points on Earth's surface that are within two degrees of latitude of Earth's Equator.

Ceres is the largest asteroid. When discovered, it was called a planet. As new asteroids were discovered, they were added to the list of planets, until some astronomy texts announced that the total number of planets had risen to 23 (as noted by Neil deGrasse Tyson, director of the Hayden Planetarium). Then, because of their small size and the great number of them that were being discovered, these small bodies were demoted from planet status to asteroids. (Pluto would be demoted for similar reasons in 2006. As a consolation prize, Pluto and Ceres are now referred to as dwarf planets.)

The width of the main belt of asteroids is shown in proper scale relative to its circumference in the map. It forms a thick belt centered on the sun. We are off center, so on the side toward the sun, the asteroid belt is farthest from us, whereas on the side opposite the sun, it is closer to us. This causes the wave appearance of the belt in the map.

The main belt asteroids lie approximately in the plane of Earth's orbit, which is tilted at an angle of 23.5 degrees relative to Earth's equatorial plane, yielding two dense clusters where the plane of the belt cuts Earth's equatorial plane and the density of asteroids is highest.

Jupiter is shown behind the sun. Beyond Jupiter are Saturn, Uranus, and Neptune.

Ever wonder what happened to Halley's comet? Here it is, between the orbits of Uranus and Neptune, as of August 12, 2003. It came close to Earth in 1835, 1910, and 1986, but it is now moving along its very elliptical orbit to its point of maximum distance from the sun. It will be back near 1 AU from the sun in 2061.

Comets are dirty snowballs made mostly of water ice—some more than ten kilometers (six miles) across.

As comets approach the sun, they become heated, and outgassing vapor gives them long visible tails—millions of kilometers long.

We find Pluto among the other icy Kuiper belt objects. We show all 771 known Kuiper belt objects (rather than just those in the equatorial plane). The band of Kuiper belt objects is relatively narrow because more distant ones would be too faint to see with current surveys. The belt shows vertical stripes of high density, depending on where various surveys were conducted in the sky. New Kuiper belt objects are being discovered all the time. Today it would be fair to call Pluto the second largest known Kuiper belt object (after Eris). Some comets originate in the Kuiper belt.

The Pioneer 10, Voyager 1, and Voyager 2 spacecraft are shown headed away from the solar system. These are on their way to the heliopause, where the solar wind (made up of particles from the sun) collides with interstellar gas.

The Oort cloud is the reservoir of long-period comets. Comets in this cloud are perturbed by passing stars onto orbits that will bring them into the inner solar system. On the way in, they are most likely significantly perturbed by Jupiter so that they either escape the solar system entirely (as happened to Caesar's comet and Tycho Brahe's comet, which are now on their way out) or are captured into a shorter-period orbit, as apparently happened to Halley's comet.

Mars at its point of closest approach in 2003 as seen by the Hubble Space Telescope (opposite). Halley's comet as seen in 1986 (below).

Beyond the Oort cloud are the stars. The ten brightest stars visible in the sky are shown with large star symbols. The nearest star, Proxima Centauri, shown with a small star symbol, is a member of the Alpha Centauri triple star system. Betelgeuse at the far right and Rigel at the far left are both in the constellation of Orion. Remember that the map is a 360-degree panorama, so the left- and right-hand borders coincide. You could tape the edges together to show the map as a cylinder.

The first radio transmission of any significant power to escape Earth's ionosphere and penetrate into space was the TV broadcast of the opening of the 1936 Berlin Olympics, a fact used in the movie *Contact*. The wavy line labeled "Radio signals from Earth have traveled this far" shows how far this first transmission had reached as of August 12, 2003. Stars below this line would have received signals from Earth. This line advances outward at a rate of one light-year per year.

Several hundred stars are now known to have planets circling them. Many of these are solar-type stars whose planets were discovered by stellar velocity variations. The planets are typically more massive than Earth but less massive than about 12 times the mass of Jupiter. As a Jupiter-size planet circles its star, it causes the star to wobble back and forth, and this causes a variation in the velocity of the star, back and forth—toward or away—from Earth, which can be detected as tiny blue- or redshifts in its spectral lines.

The star HD 209458 has a Jupiter-mass planet that was discovered by velocity variations but was later also observed in transit. Many of these planets are in tight orbits around their stars at distances of less than 1 AU because these are the easiest to detect by velocity perturbations on the star. The planet around OGLE-TR-56 was discovered by transit.

PSR 1257+12 is a pulsar (or neutron star) circled by three terrestrial-mass planets. As a neutron star rotates,

FYI Edmund Halley (1656–1742) predicted a comet he had observed in 1682 would return in 1758. When it did they named it after him.

its magnetic field lines rotate, and as the line extending from a magnetic pole sweeps by Earth, we observe a pulse of radio waves—hence, the term pulsar. The three small planets were detected by radial velocity variations of the pulsar as revealed by accurate pulse timing. It was the first star (other than our sun) discovered to have planets.

Other interesting representative objects in our galaxy are illustrated: the Pleiades open star cluster; M13, a prominent globular cluster; and the Crab Nebula, site of a supernova seen in 1054 and containing the Crab Nebula pulsar, a neutron star spinning on its axis 30 times a second. Cygnus X-1, one member of a binary star system, is a black hole about nine times as massive as the sun. The Orion Nebula (M42) and the Dumbbell Nebula are also included.

The Hulse-Taylor binary pulsar (two neutron stars in a tight binary orbit) was used to provide sensitive tests of Einstein's theory of general relativity. As the two neutron stars orbit each other, the gravitational field in the space around them is constantly changing, sending ripples—or gravity waves—in the curvature of space outward at the speed of light. This emission of gravity waves steals orbital energy

The Antennae (opposite top) are a pair of galaxies that have collided. Tidal interactions between them have sent out long tails that look like the antennae of an insect—hence the name. The distance to the M81/M82 group (opposite bottom) was determined by the Hubble Space Telescope using Cepheid variable stars.

Antennae Galaxies
75 million light-years away

M81/M82 Group
12 million light-years away

from the binary, causing the two stars to spiral slowly inward toward each other. This tiny effect, predicted by Einstein himself, was observed by Joseph Taylor and Russell Hulse, earning them the Nobel Prize in physics in 1993.

The spiral arms of the Milky Way can be seen clearly in the map. Individual arms are labeled. At the center of

FYI George Smoot and John Mather discovered tiny fluctuations in the cosmic microwave background—seeds of galaxy formation. They shared the 2006 Nobel Prize in physics.

the spiral pattern (between the Hulse-Taylor pulsar and M13 on the map) is the galactic center, which harbors a four-million-solar-mass black hole.

Beyond the Milky Way are the Large and Small Magellanic Clouds, satellite galaxies of the Milky Way. The Andromeda galaxy (M31) and the Milky Way are the two principal galaxies in the Local Group. M33, another spiral galaxy, lies nearby.

Other famous galaxies shown include the Whirlpool Galaxy (M51), a spiral, and the Sombrero galaxy, an edge-on galaxy. The galaxies M81, M87, M74, and M77 are also indicated.

The CfA Great Wall is shown in blue. Rather than plotting individual galaxies, we display the density of galaxies found along the wall as different shades of blue.

In the center of the CfA Great Wall is the Coma Cluster, one of the largest clusters of galaxies known. The Great Attractor is a concentration of mass toward which our Virgo supercluster (centered on M87) is being drawn. We still participate in the general expansion of the universe but show a small extra velocity in the direction of the Great Attractor.

The red dots appearing beyond M81 in the map are the 126,594 Sloan Digital Sky Survey galaxies and quasars in the equatorial slice (within two degrees of the celestial equator). The galaxies and quasars appear in two broad stripes. The blank regions are due to the zone of avoidance, where Earth's Equator cuts the galactic plane, and which the Sloan Survey does not cover.

On the left, at a distance of about 120 megaparsecs, we can see a large circular void in the distribution of galaxies. To the right, at a distance of 200 to 350 megaparsecs, we can see the Sloan Great Wall of galaxies. Even though the Sloan Great Wall appears about two-thirds the size of the CfA Great Wall on this map, it is actually about twice as large because, being 2.5 times farther away than the CfA Great Wall, it is plotted at a scale 2.5 times smaller.

Quasar 3C 273 is the first quasar to have its redshift measured (by Maarten Schmidt in 1963). Quasars appear to be powered by disks of gas spiraling into giant black holes in the centers of galaxies. As faster gas orbiting closer to the hole brushes against slower gas circling a bit farther out, the disk becomes heated and begins to glow brightly. The energy loss resulting from the frictional heating causes the gas to spiral slowly inward. As the gas tries to funnel in, jets of high-energy particles can squirt out at the poles, perpendicular to the plane of the disk. Quasar 3C 273 has such a jet.

The gravitational lens Quasar 0957+561 is shown, as is the lensing galaxy producing two images of the quasar. The lensing galaxy is along the same line of sight but at about one-third the distance.

A line is shown marking the epoch that divides the universe's decelerating and accelerating phases. Beyond this line (earlier in the history of the universe), the expansion of the universe is decelerating (slowing) with time. Closer than this line (in the more recent past), the expansion is accelerating because of the effects of dark energy.

The gamma-ray burster GRB 990123 is shown—for a brief period of time the brightest object in the visible universe. A number of gamma-ray bursts have been found to be associated with supernovae explosions.

The Subaru Deep Field galaxy and the gamma-ray burster GRB 090423 are also shown.

The epoch of the first stars is indicated with a line. Beyond this line, we are looking back to a time when stars have yet to form.

The most distant thing we can see is the cosmic microwave background radiation, predicted by George Gamow, Ralph Alpher, and Robert Herman in 1948. Arno Penzias and Robert Wilson discovered it in 1965—for which they earned the 1978 Nobel Prize in physics. The cosmic microwave background photons arrive directly from an epoch only 380,000 years after the big bang. Since the cosmic microwave background covers the entire sky, its multicolored line covers the full 360-degree width of the map at the top.

We have learned much by studying the small temperature fluctuations in the cosmic microwave background, which have been measured with high accuracy and analyzed by the WMAP satellite team of Lyman Page, David Spergel, and Charles Bennett and have led to the accurate cosmological model that allowed us to draw this map. The pattern of fluctuations we see—shown on a scale varying from blue to green to red—matches the pattern predicted by Alan Guth's 1981 theory of inflation, which says that the very early universe started off as a tiny region containing a lot of "dark energy." This dark energy had a very large repulsive gravitational effect, creating the big bang explosion.

Maps can change the way we look at the world. Gerardus Mercator's cylindrical map projection, first presented

in 1569, was influential not only because it showed the shapes of continents well but also because, for the first time, fairly accurate contours for North and South America could be shown. Now that astronomers have arrived at a new understanding of the universe, all the way from the solar system to the cosmic microwave background, our Map of the Universe should provide a new visual representation of these exciting discoveries.

The WMAP satellite map of the cosmic microwave background sphere. We on Earth observe from a central position inside so it spans 360 degrees on the Map of the Universe.

SIZES IN THE

CHAPTER 5

SOLAR SYSTEM

TERRESTRIAL PLANETS

NOW THAT WE KNOW the angular sizes of objects from Chapter 1 and their distances from Chapter 3 as plotted on the Map of the Universe in Chapter 4, we can determine their sizes. Now we can give the answers. We will start in our own backyard—with objects in the solar system. First are the terrestrial planets, the four rocky planets nearest the sun: Mercury, Venus, Earth, and Mars. On the following spread we show them all at the same scale of 1:100,000,000. At this scale, 1/16 of an inch equals 100 miles.

FYI Among the terrestrial planets in our solar system, Venus is the closest to the size of Earth—only about **5** percent smaller.

Mercury, with a diameter of 4,880 kilometers (3,030 miles), is the smallest bona fide planet. It has a heavily cratered surface similar to that found on the moon. Mercury has more than 60 craters 150 kilometers (90 miles) across or larger. Most of this cratering took place during a period of heavy bombardment 3.8 billion years ago. Mercury has a large iron core, giving it a density almost as high as that of Earth. Two icy moons in the solar system (Jupiter's moon Ganymede and Saturn's moon Titan) are slightly larger, but no moon is as massive as Mercury.

It has a trace atmosphere—composed primarily of oxygen, sodium, helium, potassium, and hydrogen—with less than one-trillionth the atmospheric pressure found on Earth. Without a thick atmosphere to regulate temperatures, its surface varies from minus 170°C (-274°F), on the night side before sunrise, to 430°C

(806°F) at noon. The tidal gravitational effects of the sun lock Mercury's rotation period to its orbital revolution period around the sun such that Mercury rotates on its axis three times for every two orbits. Mercury rotates on its axis once every 59 Earth days, and it takes 88 Earth days to circle the sun.

Venus has a diameter of 12,104 kilometers (7,521 miles), almost as large as Earth's, so it is often called our sister planet. Its thick carbon dioxide atmosphere exerts a crushing pressure on its surface: 92 times that found on Earth. This produces a tremendous greenhouse effect, raising surface temperatures to 460°C (860°F). Venus is completely shrouded by a sulfuric acid cloud layer stretching in altitude from 45 kilometers to 70 kilometers (28 miles to 43 miles), blocking our view of its surface.

However, radar mapping by the Magellan spacecraft has revealed the surface topography in great detail, as shown on pages 140-141. Venus does not have plate tectonics, so heat from its interior builds up until the crust cracks and lava flows out, resurfacing the entire planet. The last such event occurred about 500 million years ago. Venus shows lowland, hilly, and highland regions, plus many volcanoes and craters. The total altitude variation on the surface from lowest to highest point is 11 kilometers, or 7 miles (about half the variation seen on Earth). A "day" on Venus (from noon to noon) takes 116 Earth days. The sun rises in the west and sets in the east. There are 1.92 such Venus days in a Venus year.

Earth has a diameter of 12,756 kilometers (7,926 miles). Seventy-one percent of Earth's surface is covered by oceans, with a mean water depth of 3.7 kilometers (2.3

between day and night. The moon has 70 craters measuring 150 kilometers in diameter (90 miles) or larger. Average surface temperatures range from 130°C (260°F) in the day to minus 180°C (-290°F) at night.

Mars has a diameter of 6,760 kilometers (4,200 miles). Its atmosphere is composed primarily of carbon dioxide, with an average surface pressure 0.6 percent of that found on Earth. Mars has two tiny moons, Deimos and Phobos, whose diameters are 15 kilometers and 27 kilometers respectively (9 and 17 miles). On our picture at 1:100,000,000 scale, these have diameters of 0.15 millimeter and 0.27 millimeter (0.006 in. and 0.011 in.). We have actually plotted them in the picture, at those sizes, but they are at the limit of the printing resolution, so they are not really visible.

This completes the terrestrial planets—their group portrait follows.

miles). The mean elevation of the land area is 0.8 kilometer (0.5 mile). Earth's atmosphere is composed primarily of nitrogen and oxygen. Surface temperatures range from a record high of 58°C (136°F) to a record low of minus 89°C (-128°F). Earth has a rich and diverse biosphere with millions of species, including *Homo sapiens*, a species that has measured Earth's size, discovered its place in the universe, and traveled to visit Earth's moon.

Earth's moon has a diameter of 3,476 kilometers (2,160 miles). It is tidally locked so that it always keeps the same face toward Earth, making its rotation period on its axis the same as its orbital period circling Earth. Thus, a "day" on the moon lasts 29.5 Earth days, the same as the time from full moon to full moon. The moon, just like Mercury, has only a trace atmosphere, not enough to stop meteors or regulate its surface temperatures

Cloud-shrouded Venus, one of our solar system's four terrestrial planets. This image was taken from the Mariner spacecraft.

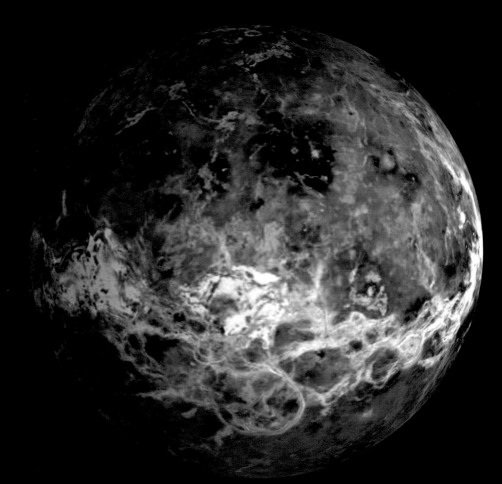

Mercury

Venus

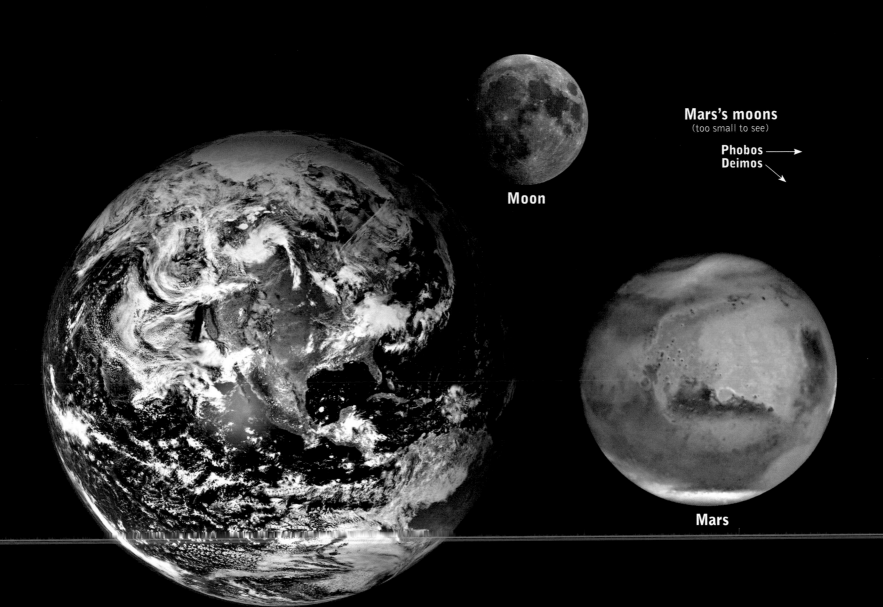

Moon

Mars's moons
(too small to see)

Phobos ⟶
Deimos
↘

Mars

Earth

Earth

IN ADDITION TO BEING covered with water, extensive areas of vegetation are visible on the land areas. Earth has a distinctive spectrum that we may use as a guide when looking for Earthlike planets orbiting other stars. The highest altitude point on Earth's surface is Mount Everest (+8,848 meters above sea level). It was first reached by Edmund Hillary and Tenzing Norgay in 1953. The lowest altitude point on Earth is at the bottom of the Mariana Trench in the Pacific Ocean (minus 11,033 meters). It was first reached by Don Walsh and Jacques Piccard in the U.S. Navy bathyscaphe submarine *Trieste* in 1960.

FYI Earth has 13.4 times as much surface area as the moon.

The maps shown here are both at the same scale. They are plotted using an Aitoff-Wagner projection in a version Gott designed to minimize overall distortions by depicting shapes and areas as accurately as possible while minimizing bending, lopsidedness, and length of boundary cuts, using principles developed in collaboration with David Goldberg. Earth, being larger than Mars or the moon, has retained an appreciable atmosphere yet does not suffer from an excessive greenhouse effect like Venus. It also has an ozone layer protecting us from UV radiation from the sun.

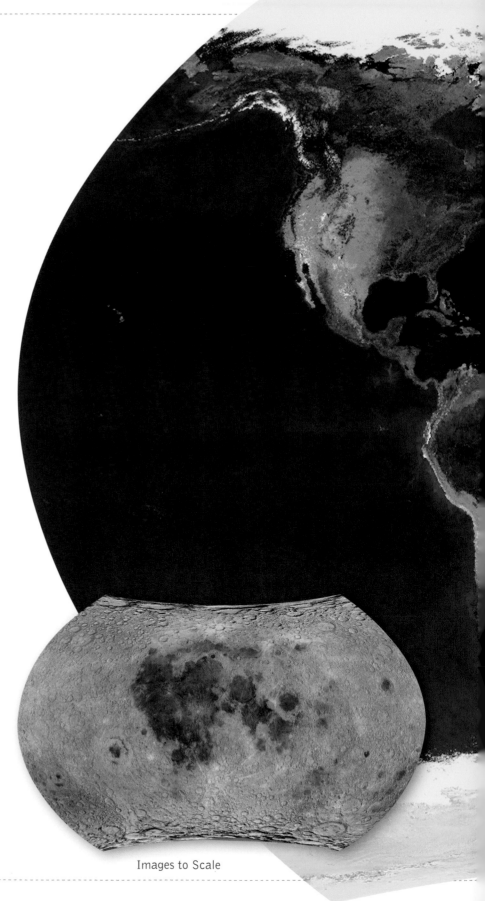

Images to Scale

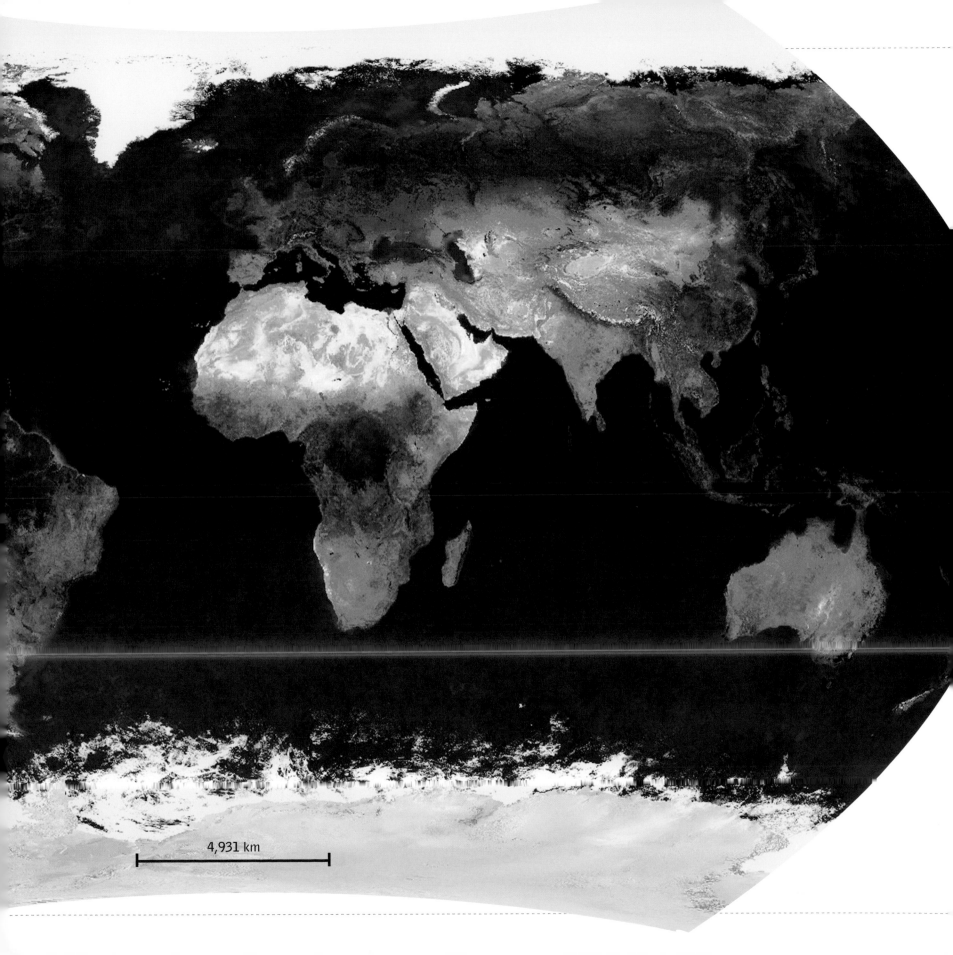

4,931 km

Earth in Perspective

THE MOON IS ROUGHLY a quarter the size of Earth. So when astronauts on the moon looked back at Earth, it was four times the angular diameter of the moon as seen from Earth. All the Apollo astronauts marveled at this spectacular sight. We show pictures of Earth taken from the moon, and the moon taken from Earth, to scale, so you can compare the views. To see them in proper angular scale, view the pair of pictures from a distance of 100 inches. Finally, there is a picture of the moon passing nearly in front of Earth taken by the Voyager spacecraft.

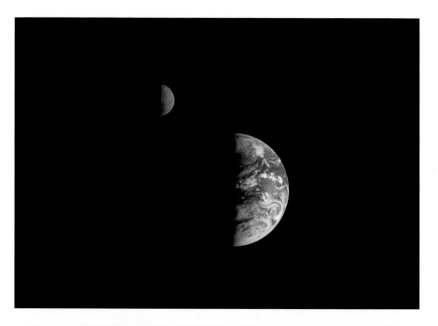

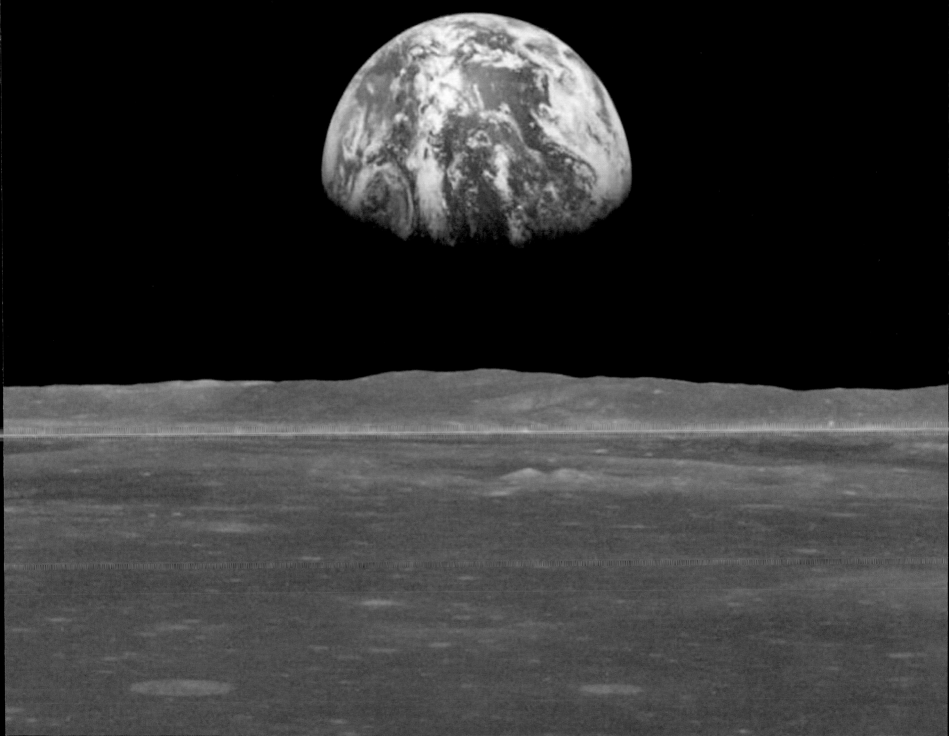

Moon

AT LEFT, THE VISIBLE and back faces of the moon are shown as hemispheres. Opposite is a map of the entire moon using the Gott-Mugnolo azimuthal map projection, which has the smallest distance errors. We show Earth using the same projection, below. The large map shows the visible hemisphere of the moon in the center but allows us to peek around the edges to see the back side. The north and south poles of the moon, where the sun always strikes at a low angle, show appreciable shadows.

Mare Imbrium is a large impact basin formed about 3.85 billion years ago. It and other dark maria, or seas, are actually basins that later filled with basalt lava flows, which then solidified. Most of the maria are found on the near side of the moon, where the crust is thinner.

The Copernicus crater was formed 900 million years ago. The asteroid that impacted to form the Tycho crater 108 million years ago appears to have been part of a swarm of asteroids created when a large asteroid was shattered by a collision in the asteroid belt some 160 million years ago.

Earth shown using same projection

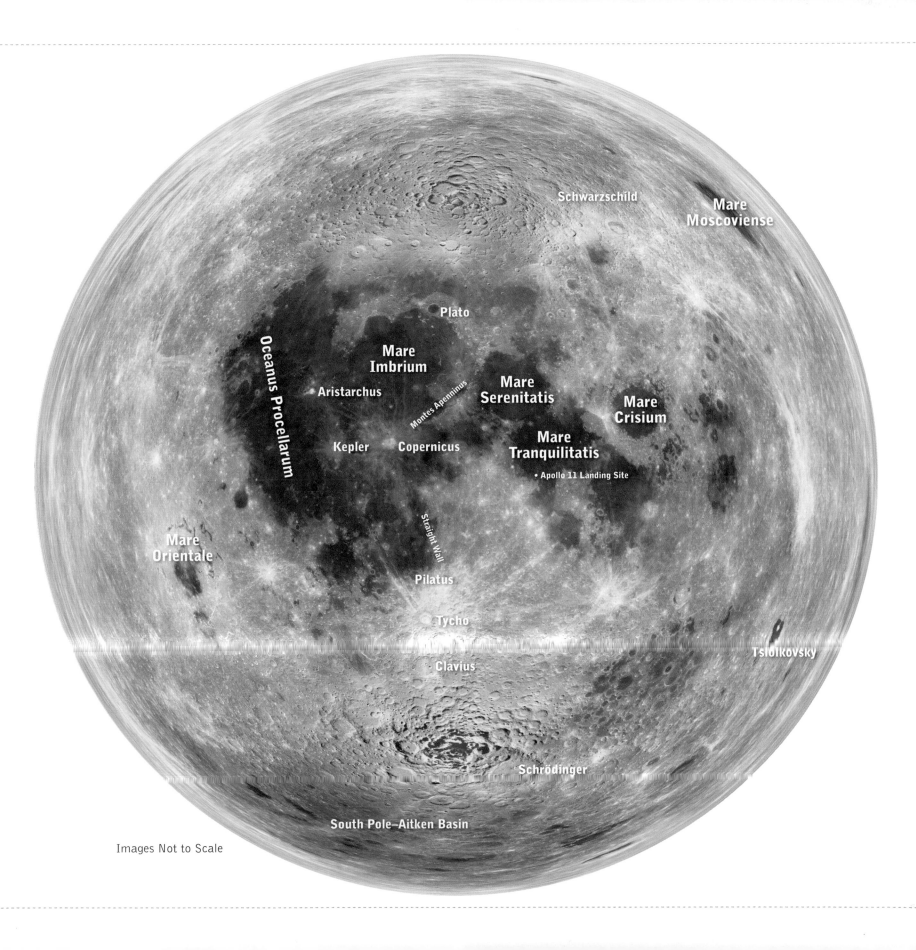

Schwarzschild

Mare Moscoviense

Plato

Oceanus Procellarum

Mare Imbrium

Aristarchus

Montes Apenninus

Mare Serenitatis

Mare Crisium

Kepler

Copernicus

Mare Tranquilitatis

• Apollo 11 Landing Site

Mare Orientale

Straight Wall

Pilatus

Tycho

Tsiolkovsky

Clavius

Schrödinger

South Pole–Aitken Basin

Images Not to Scale

Copernicus Crater

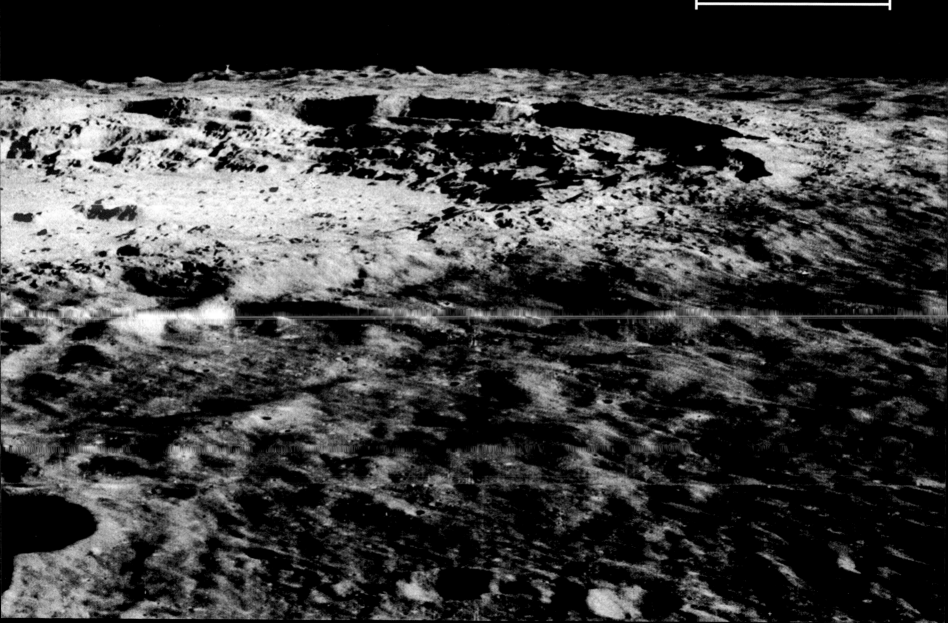

21.1 km

The Lunar Landscape

THIS 360-DEGREE lunar panorama is from the Apollo 15 mission next to the Apennine Mountain range on the southeastern edge of Mare Imbrium. It gives you a feeling of what it is like to stand on the moon and look around.

The topography is smoothed by billions of years of bombardment by micrometeorites. The highest altitude point on the moon is not a mountain at all, but a crater rim on the lunar highlands on the far side of the moon with an altitude of 10,750 meters (35,270 feet). The low point on the moon is minus 9,060 meters, at the bottom of the South Pole–Aitkin Basin, an ancient giant impact basin also located on the far side. This compares with the plus-8,848-to-minus-11,033-meter range found on Earth—from Mount Everest to the Mariana Trench. The altitude variation on the moon is about the same as that found on Earth, even though it is much smaller than Earth.

Astronaut David Scott is shown next to the lunar rover facing the summit of smaller and nearer Mount Hadley Delta. The rover allowed the astronauts to venture more than four kilometers (two miles) from their landing site. James Irwin took the picture—you can see his shadow on the ground, just 180 degrees (or halfway around the 360-degree panorama) from the sun.

I was once on a panel with Harrison Schmitt, one of the last two astronauts to walk on the moon. Given a unique chance to talk to someone who had actually walked on the moon, I asked him what surprised him about the mission—what didn't the training simulations capture?

He said he was surprised to see a meteor passing below him while he was orbiting Earth. He was also taken by surprise when flying over the dark portion of the moon illuminated only by earthshine. He saw beautiful blue craters

passing under him as he orbited. (Earthshine, light from Earth, is mostly blue because of Earth's oceans.) When he was on the surface of the moon, he said that it was hard to judge distances because you are in the vacuum of space, with no atmospheric haze.

I also asked him how big Earth looked from the moon. Because it was way up in the sky, with nothing to compare it with, he said it was hard to tell. Of course, a camera would have told you it was 3.66 times bigger in angular size than the moon looks from Earth.

Earth is also high in the sky from the Mount Hadley region—about 65 degrees above the horizon—that's why you don't see it in this panorama, which hugs the horizon. The same side of the moon always faces Earth as it circles the planet, so at any given location, Earth stays in approximately the same position in the sky all the time.

Since Earth is one of the most spectacular sites you can see from the moon, the best location for a tourist hotel on the moon would be on the near side near the north pole. That way Earth would always hang just above the horizon, where it could be seen against distant lunar terrain and would look largest. And since you would be near the moon's north pole, Earth would be right side up as you looked at it—with its North Pole at the top. Evidence for water ice (at about one part per thousand) was found in lunar soil blasted from a permanently shadowed crater near the south pole. We suspect similar water ice deposits exist near the north pole.

Mount Hadley, the big mountain with a prominent shadow just to the left of the sun, 4,572 meters from base to peak, is the second tallest mountain on the moon. It is almost 30 kilometers away.

Mars

THIS GOTT-COLLEY MAP OF MARS uses the Gott elliptical projection. Equal areas on the map correspond to equal areas on the surface of Mars. I designed this so that shapes are perfect along the central meridian. This projection is useful for Mars, because it shows Mars's important polar caps well. They are not stretched out as they would be on a Mollweide equal-area projection.

The highest point on Mars (+21,281 meters) is Olympus Mons, a giant extinct volcano over twice as tall as Mount Everest. On Earth, a magma hot spot created a chain of island volcanoes (Hawaii) as the Pacific plate moved over it. But on Mars, a smaller planet, plate tectonics ended early, so a magma hot spot created one enormous volcano.

Valles Marineris is a giant rift canyon 4,000 kilometers long, 600 kilometers wide, and 6 kilometers deep. The polar caps and Syrtis Major, the most prominent dark (basalt rock) area, can be seen with a small telescope when Mars makes a close approach to Earth once every two years. Between Valles Marineris and Terra Meridiani are sinuous valleys, clearly indicating that liquid water once flowed there, emptying northward.

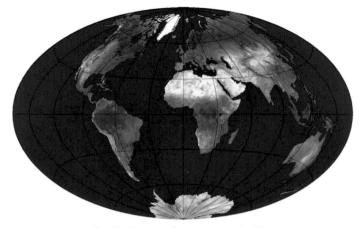

Earth shown using same projection

Tempe
Terra

Olympus Mons

Tharis Montes

Valles Marineris

Solis Planum

Terra
Sirenum

Images Not to Scale

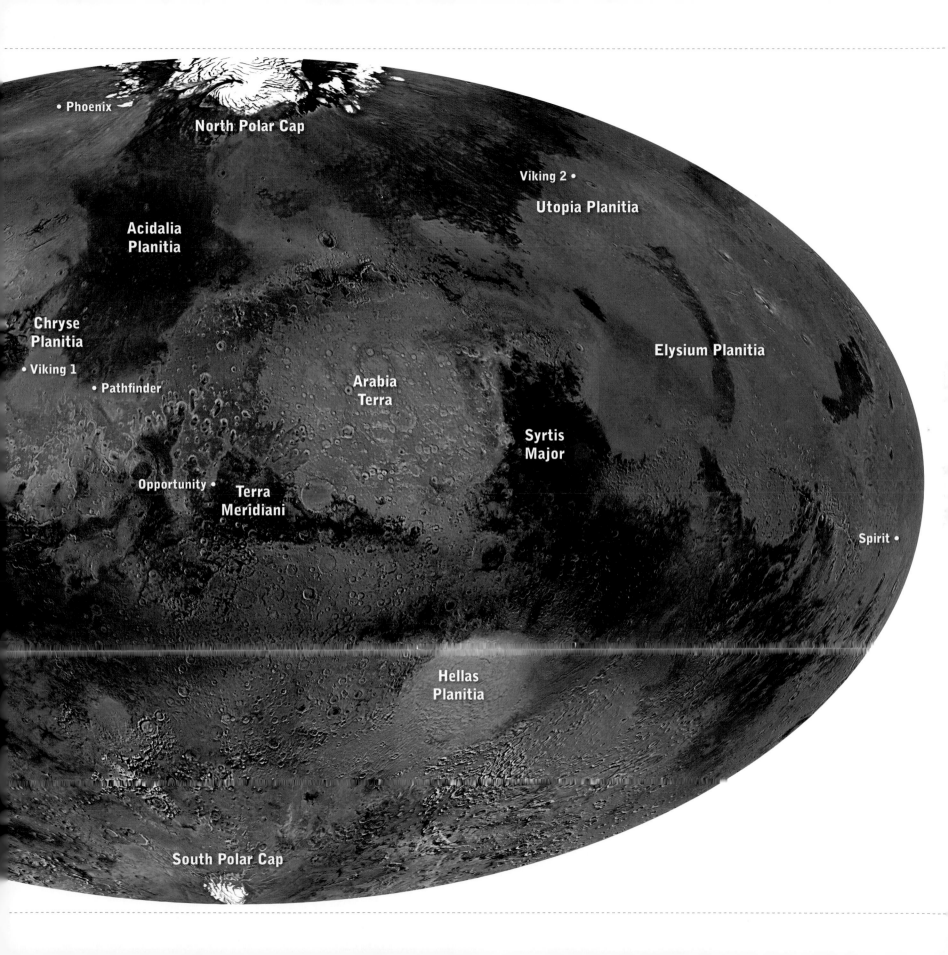

■CANYONS COMPARED

Valles Marineris

Grand Canyon

189 km

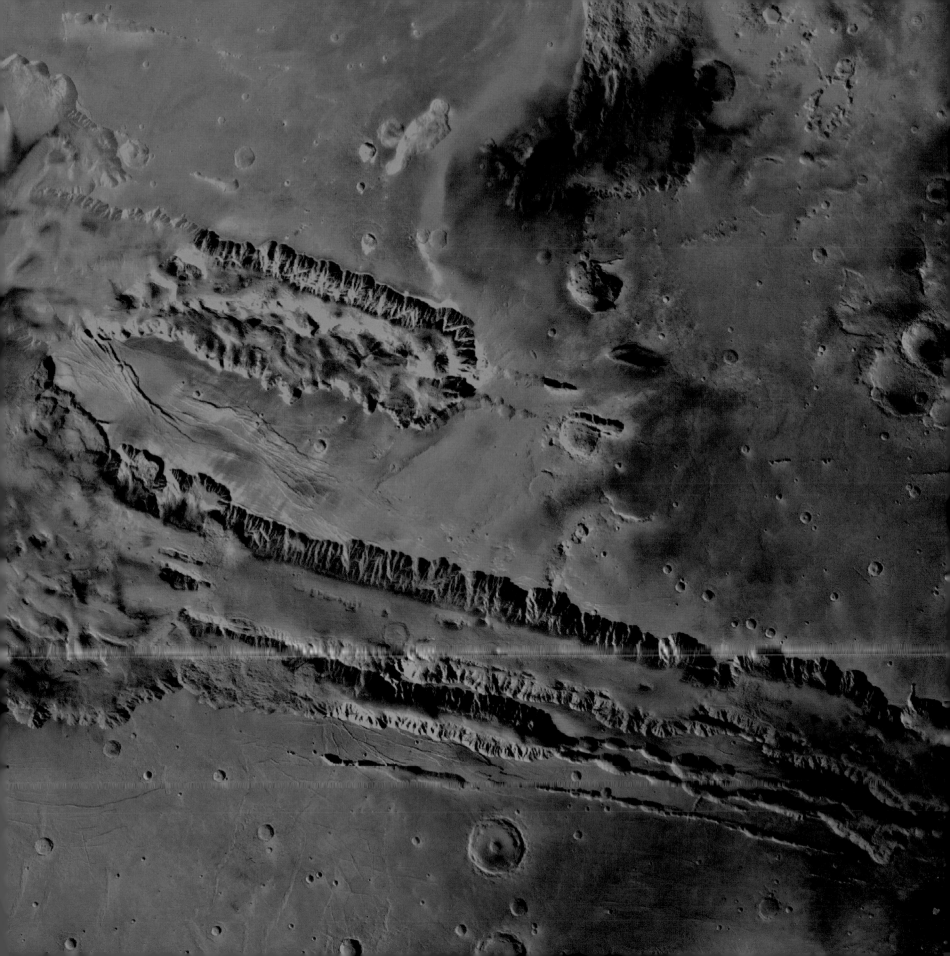

The Martian Landscape

THIS IS A 360-DEGREE panorama from the Mars Pathfinder lander. A ramp was extended, and the small rover, Sojourner, moved off onto the surface of Mars.

The view here is more colorful than the lunar panorama. The landing was made in a region of sinuous valleys indicating liquid water had once flowed copiously there. The flood plain visible in this panorama is covered by boulders swept along by these floodwaters.

The Sojourner rover appears in the picture right next to a prominent boulder fondly named Yogi. The rover has six wheels and solar cells on its top to provide power. It is equipped with an Alpha Proton X-ray Spectrometer that can be placed on a rock surface. The spectrometer emits alpha particles (helium nuclei) produced by radioactive decay. These particles slam into the rock surface and produce x-rays, protons, and backscattered alpha particles. By analyzing these particles, one can determine the elemental composition of the rock, which enables geologists to suggest plausible mineral compositions. These results suggest that Yogi is made mostly of feldspar, quartz, and orthopyroxene. Yogi is about one meter tall and about five meters away from the lander.

The twin peaks in the background are only about 50 meters (164 feet) high and about 1.0 kilometer (0.6 mile) away. They look like much more distant mountains. Distances can be deceiving. As this book tells you, you have to know the distance to something to go from its apparent to its actual size. The twin peaks were located on pictures from orbit, along with other landmarks in the panorama. By triangulation one could locate the position of the lander and determine the distance to the twin peaks. Knowing that, the picture tells us their height.

Our landers and rovers have delivered much information about Mars. The Phoenix lander near the north polar cap found permafrost. Its panoramic view showed the characteristic polygonal pattern found in

permafrost arctic terrain on Earth. A scoop yielded water ice. The Opportunity rover found "blueberries"—hematite-enriched concretions—indicating past liquid water, as well as the mineral jarosite, which forms only in the presence of water.

Mars has much more water than the moon. The planet has polar caps of water ice as well as permafrost extending outward from the polar areas. Mars's north polar cap has a volume of 1.6 million cubic kilometers (384,000 cubic miles), about half that of the Greenland ice cap on Earth. The fraction of Martian soil composed of water ice approaches 100 percent as one reaches the poles. In contrast, as one approaches the poles on the moon, this fraction is only about one part in one thousand.

Liquid water has also occasionally been seen on the Martian surface today, in flows released from the permafrost. Mars, unlike the moon, also possesses plentiful amounts of elements such as carbon, nitrogen, potassium, and sodium, which are necessary for life as we know it. Mars is particularly interesting scientifically both as a comparative case in planetary climatology and as a promising place to search for evidence of either present or fossil microscopic life.

As a site for space colonization, Mars is promising. It has all the chemicals necessary for life, reasonable gravity (38 percent that of Earth), an atmosphere, and temperatures that are less extreme than other possible sites in the solar system. Caves with entrances more than 100 meters (330 feet) across have even been seen, perhaps furnishing sites where one could hide out from galactic cosmic rays and solar particle events.

This is a 360-degree panorama from the Mars Pathfinder lander. The viewpoint is about 1.5 meters (5 feet) above ground level. After landing, the Pathfinder lander was renamed the Carl Sagan Memorial Station, in honor of astronomer Carl Sagan.

Sunsets Compared

MARS IS FARTHER from the sun than Earth, so the sun looks smaller from Mars. On Earth, atmospheric bending of light (refraction) compresses the sun's image as it sets, making it slightly elliptical. Mars's atmosphere is too thin to cause such an effect. Scattering of blue light in Earth's atmosphere causes the setting sun to appear orange, but on Mars the atmosphere is thin and atmospheric scattering is less pronounced, so the sun from Mars in this picture shows its true white color even as it sinks below the horizon.

One objective of Mars exploration is the search for any evidence of possible life—past or present. Some liquid water has been found flowing on Mars in short-lived flows released from subsurface permafrost, and water ice has been found in permafrost near the north polar cap. We know that liquid water flowed freely on Mars early in its history. If microbial life developed, then it might have gone extinct as Mars dried out—so a search for microfossils is indicated. We might even search for present subsurface life.

A day on Earth lasts 24 hours, but a day on Mars lasts 24 hours, 39.6 minutes. Mars would be a good place for night owls, whose biorhythms make them want to stay up late at night and sleep later in the morning. The temperature ranges from minus 140°C (-220°F) on the polar caps in midwinter to a high of 20°C (68°F) at some locations in summer. The atmosphere is 95 percent carbon dioxide at about one percent the atmospheric pressure we have on Earth. Space suits would be required!

A primary danger to any Mars colonist would be the increased risk of cancer because of galactic cosmic rays (high-energy atomic nuclei) and solar particle events. From this standpoint, the safest place to plant a Mars colony would be Hellas, the lowest point on Mars, where the atmospheric protection is the greatest. There would likely be ready access to water ice in permafrost there. If the colony were built 10 meters (33 feet) underground and the colonists spent 30 hours a week on the surface, they could still have a life expectancy of 70 years.

Any colonists on the surface could see interesting transits (with suitable eye protection). Mars's moons Deimos and Phobos transit in front of the sun periodically. We have pictures of such transits from our Mars rovers.

In 1972, Arthur C. Clarke wrote a science fiction story about an astronaut on Mars who saw a transit of Earth and moon in front of the sun occurring on May 11, 1984. He thought astronauts might be on Mars by 1984 to see it. The next transit of Earth and moon in front of the sun as seen from Mars will occur on November 10, 2084. Will there be any people on Mars to see it?

A sunset from Earth (opposite) as compared with a Martian sunset picture taken by Spirit rover at Gusev crater on May 19, 2005 (above). To see both sunsets at their correct angular sizes, view these pages from a distance of 48 inches.

GAS GIANTS

THE GAS GIANT PLANETS—Jupiter, Saturn, Uranus, and Neptune—are denizens of the outer solar system. Like the rocky terrestrial planets, they were formed when planetesimals in the early solar system collided and coalesced into larger bodies. However, in the outer solar system, where temperatures are much lower, such rocky cores may also accumulate large envelopes of hydrogen and helium gas, forming gas giant planets.

Our size comparisons in the next two-page spread are shown at a scale of 1:1 billion, where 1/16 of an inch equals 1,000 miles. The scale is a factor of 10 smaller than the portrait of the terrestrial planets with which we began this chapter. The gas giant planets are roughly a factor of 10 larger than the terrestrial planets; hence we need to show them at a correspondingly smaller scale. Earth is shown for comparison.

Jupiter's diameter is 142,984 kilometers (88,846 miles), and its composition is similar to that of the sun. It is mostly molecular hydrogen and helium transitioning to metallic hydrogen and helium as pressure builds toward the center, with a small icy and rocky core at the very center. Jupiter has a diameter 11 times larger than Earth's. Its equatorial diameter is larger than its diameter from pole to pole. Isaac Newton figured out that this is due to Jupiter's rapid rotation and the effect of centrifugal force holding it out at the equator. Jupiter rotates on its axis once every 9.925 hours. A day on Jupiter passes quickly.

Jupiter has a mass 318 times that of Earth. Since Jupiter is gaseous, we define its diameter by the place where its atmospheric pressure equals that found at the surface of Earth. The gravity at that level is 2.6 times that found on the surface of Earth. If you weighed 200 pounds on Earth, you would weigh 520 pounds on Jupiter. If you are watching your weight, don't go to Jupiter!

Saturn has a diameter of 120,536 kilometers (74,898 miles), with a mass 95 times as large as Earth's. It rotates on its axis once every 10.65 hours. It is also slightly flattened because of its rotation.

Saturn's composition and structure are similar to Jupiter's. It is less dense than Jupiter because its smaller mass causes less compression. Saturn is surrounded by beautiful rings made up of icy particles orbiting it.

FYI Saturn's average density is **69** percent that of water. If you could find a big enough ocean to put it in, Saturn would float!

Uranus has a diameter of 51,118 kilometers (31,763 miles). It has a hydrogen, helium, and methane envelope; a layer of hydronium, ammonium, and hydroxide molecules; and a large, rocky core at the center (occupying about 30 percent the diameter of the planet). Uranus rotates about its axis once every 17.24 hours, but its axis of rotation is tilted at 98 degrees relative to the plane of its orbit. This large tilt suggests that Uranus suffered a giant collision in the past that knocked it off its original axis. Uranus has a ring system, which is visible.

Neptune has a diameter of 49,528 kilometers (30,775 miles). Its structure is similar to that of Uranus. It rotates on its axis once every 16.11 hours. Its blue spot, like Jupiter's red spot, is a storm in its atmosphere.

Finally, as part of our group portrait of the gas giants, we show the sun, an even larger ball of gas, peeking in at the bottom. The SOHO satellite image shows the

hydrogen—the easiest nuclear fuel to burn) and also create a nuclear furnace. Below 13 times the mass of Jupiter are the planets.

The sun rotates on its axis once every 25 days at the equator and once every 33 days at the poles. This is possible because the sun is not a solid, as Earth is. The sun is composed mostly of hydrogen; the next most common elements by number relative to hydrogen are helium (10 percent), oxygen (0.074 percent), carbon (0.035 percent), neon (0.012 percent), nitrogen (0.0093 percent), magnesium (0.0038 percent), silicon (0.0035 percent), and iron (0.0032 percent). The composition of the solar nebula from which the planets were made is similar.

photosphere, the bright surface we see in visible light. The diameter of the sun is 1,391,900 kilometers (864,887 miles), more than 100 times larger than Earth's. A million Earths would fit inside it. The sun's mass is 333,000 times that of Earth and 1,048 times that of Jupiter.

The average density in the sun is about one-fourth that of Earth. It is made wholly of gas rather than rock. Gravity compresses the gas in the sun until its interior gets hot enough to ignite nuclear reactions in its core. Hydrogen is burned into helium. It's the same energy source used in the hydrogen bomb. The temperature in the center of the sun is approximately 15.5 million kelvin. Its surface temperature is 5800 kelvin.

A star must be at least 80 times the mass of Jupiter for gravitational compression at its center to ignite hydrogen. Between a mass of 13 times to 80 times that of Jupiter, brown dwarfs are able to burn deuterium (heavy

The sun's hot, thin outer atmosphere, the corona, can be seen during a total solar eclipse when the bright photosphere is blocked by the moon. The next total solar eclipses visible from the United States occur on August 21, 2017, and on April 8, 2024.

Earth for comparison

Uranus

Jupiter

50,800 km

Saturn

Neptune

Sun

Sun

THE SUN ROTATES FASTER at its equator than at its poles, winding up lines of magnetic field in its interior like tight rubber bands. Often, a loop of magnetic field lines will pop out of the surface, creating a pair of sunspots where the loop intersects the surface. In the picture opposite, which is a SOHO satellite picture taken in ordinary visible light, you can see a number of sunspot pairs. Sunspots have a typical temperature of 4400 kelvin, compared with the average surface temperature of 5800 kelvin. Being colder, they shine less brightly than their surroundings and so appear dark by comparison. A spot's temperature must be less than its surroundings, so its lower thermal pressure plus its magnetic pressure (from its intense magnetic field) will be in balance with the higher thermal pressure outside. The mottled brighter-than-average areas near some of the sunspots are called faculae. The solar picture on the left was taken by one of us (Vanderbei) using red hydrogen-alpha light (emitted by atomic hydrogen) with a wavelength of 6,563 angstroms. This filter blocks out other wavelengths and allows us to see several large prominences (ejected gas following magnetic field lines) extending out from the edge of the sun.

FYI Never look directly at the sun. These pictures were taken with appropriate filters that cut out most of the light.

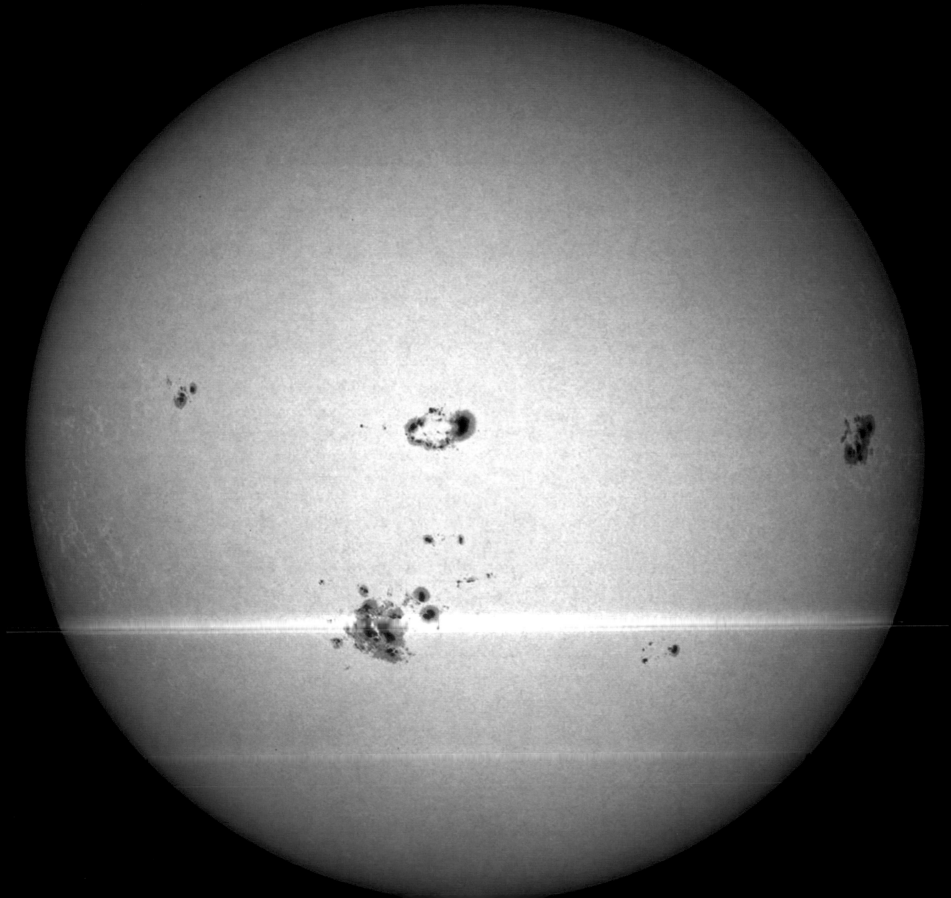

■SUNSPOT COMPARED

U.S.A.

4,330 km

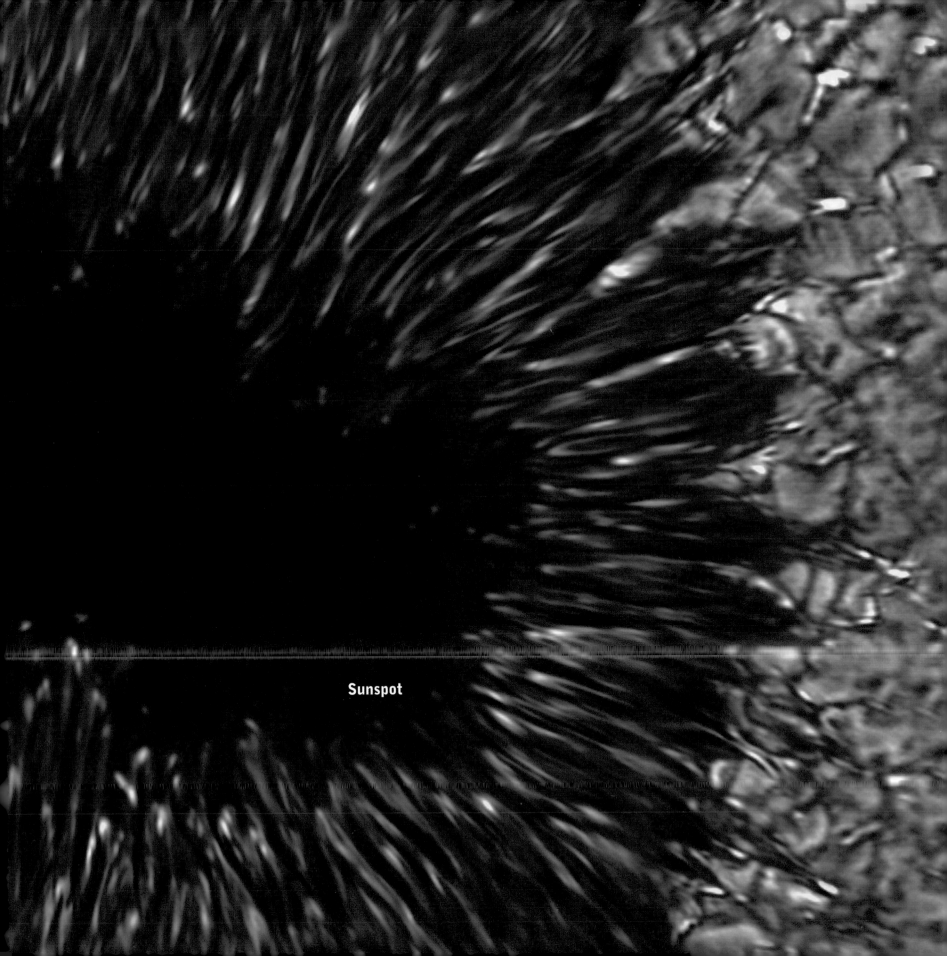

Sunspot

Jupiter

THIS MAP OF JUPITER and the corresponding reference map of the Earth, below, use a Kavrayskiy VII projection. Gott and David Goldberg found this projection to have the least overall distortion among projections showing latitude lines as straight lines: It produced the best local shapes and areas, with the least bending, lopsidedness, and boundary cuts. Straight latitude lines are important in representing Jupiter since it has cloud belts that follow latitude lines. The north pole is the line at the top of the map, and the south pole is the line at the bottom. The latitude scale is linear between poles.

Jupiter's outer atmosphere appears to have three layers of clouds: a top layer of white ammonia clouds, a lower layer of red or brown ammonium hydrosulfide, and below that a layer of water clouds, usually not visible. The darker bands are called belts, while the lighter bands are called zones. The white Equatorial Zone and the dark North Equatorial Belt and South Equatorial Belt are usually easily visible in a small telescope. The Great Red Spot is a several-hundred-year-old storm. The Red Spot rotates counterclockwise, trapped between the South Equatorial Belt and the South Tropical Zone, which pass it in opposite directions on each side.

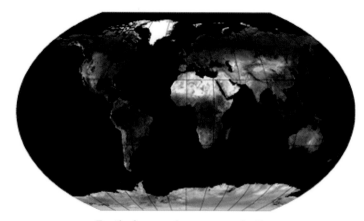

Earth shown using same projection

Images Not to Scale

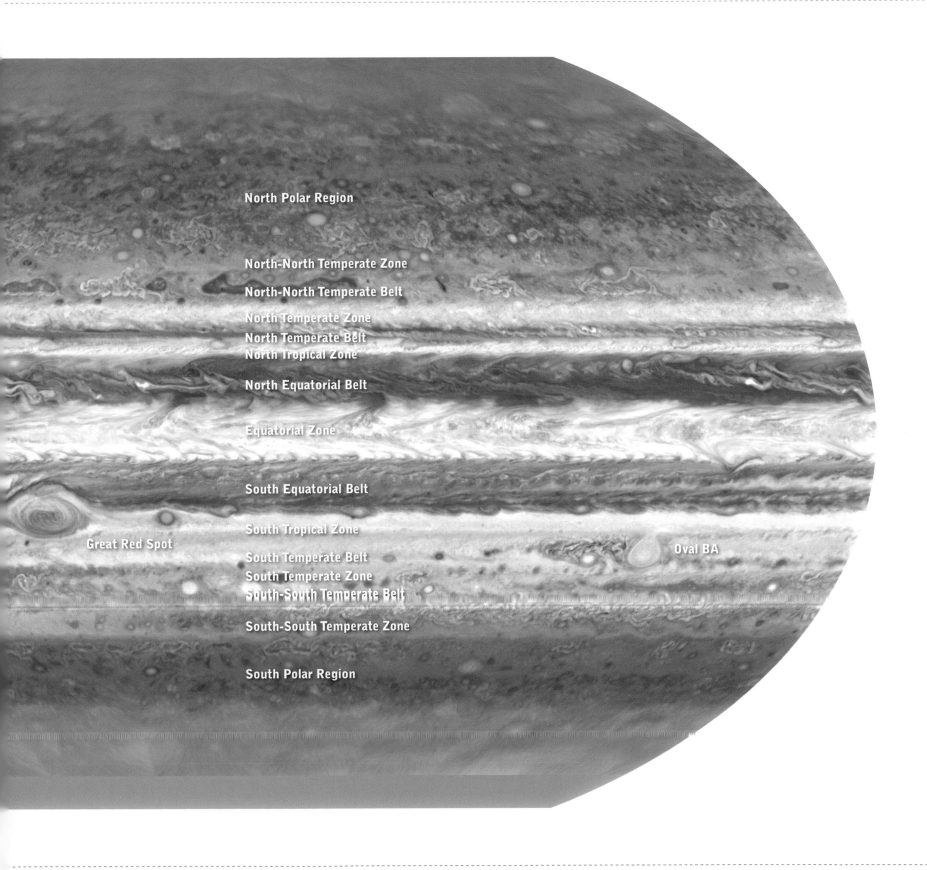

■STORMS COMPARED

Hurricane Katrina

|——————————| 4,370 km

Great Red Spot

Saturn

SATURN HAS CLOUD BANDS like Jupiter. In the picture at right, auroras (blue) similar to those we see in the ionosphere on Earth can be seen circling the north pole. (Auroras are atmospheric glows caused by interaction with solar wind particles, which, guided by magnetic field lines, impinge on polar areas.) The following two-page spread shows a close-up of an aurora on Saturn compared in size with an aurora surrounding the South Pole of the Earth. In this infrared picture of Saturn, you can see beneath the aurora a distinctive hexagonal cloud pattern surrounding Saturn's north pole, 25,000 kilometers (15,000 miles) across.

Saturn's rings are easily seen through a small telescope. Galileo glimpsed them in 1610, but to him they looked like two blobs on either side of Saturn. As Saturn circles the sun, we view Saturn and its ring system at different angles. Eventually, the rings appeared edge-on to Galileo and disappeared. Galileo was puzzled and didn't report his results. In 1655, Christiaan Huygens, observing with a better telescope, first correctly described Saturn as surrounded by a thin, flat ring, nowhere touching. The dark, narrow gap between the rings, the Cassini division, was discovered by Giovanni Cassini in 1675. In 1859, James Clerk Maxwell (1831–1879) showed that the rings could not be solid but must be composed of myriad orbiting particles.

Today we know that these particles are made of water ice plus small amounts of dust and range up to the size of houses. Particles orbiting at the inner edge of the Cassini division are in a two-to-one resonance with Saturn's moon Mimas, circling Saturn twice every time Mimas circles once. Periodic perturbations by Mimas cause particles orbiting just outside this radius to spiral outward, partially clearing the Cassini division of particles and making it

appear dark. In the picture below, we can see clearly the A ring (outside the Cassini division), the white B ring (inside the Cassini division), and inside that, the translucent C, or Crepe, ring, which can be seen against Saturn's disk.

Saturn's rings as seen from above—with Saturn's image removed so that we see just the rings (opposite). Saturn's shadow appears to the right. The sun illuminates the ring plane from below, hence we are seeing them backlit.

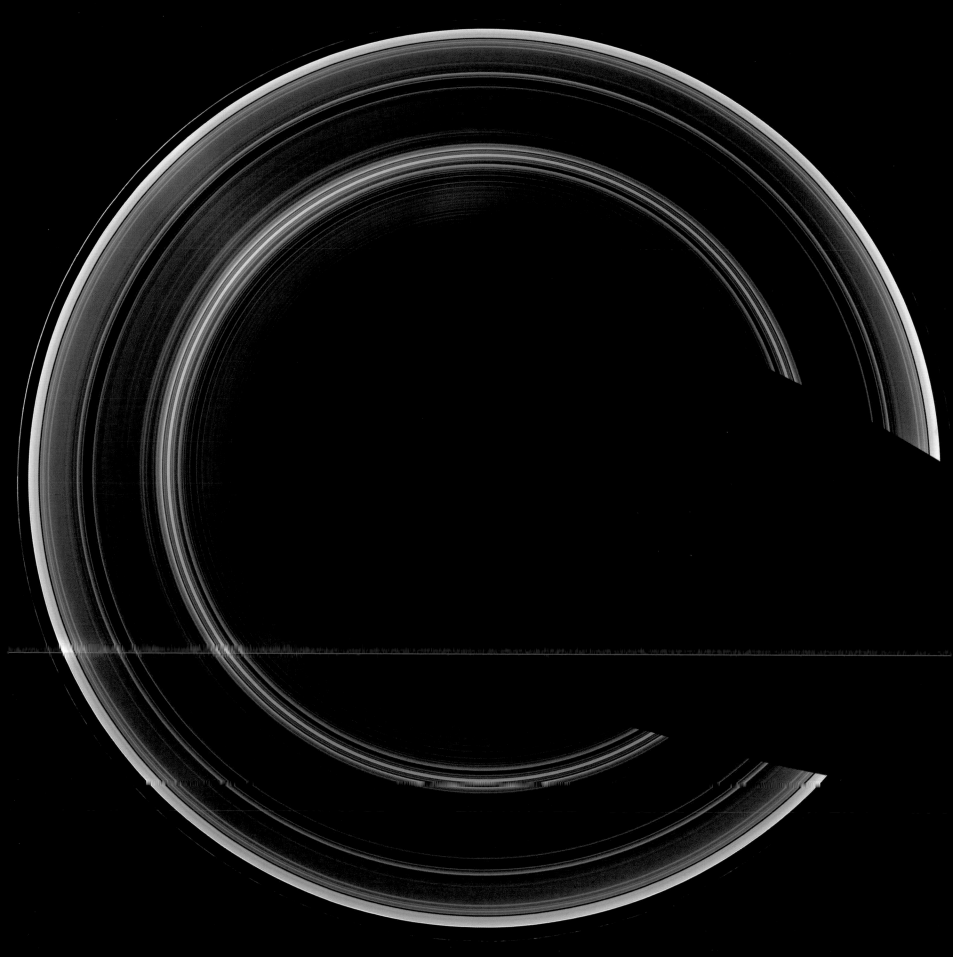

■AURORAS COMPARED

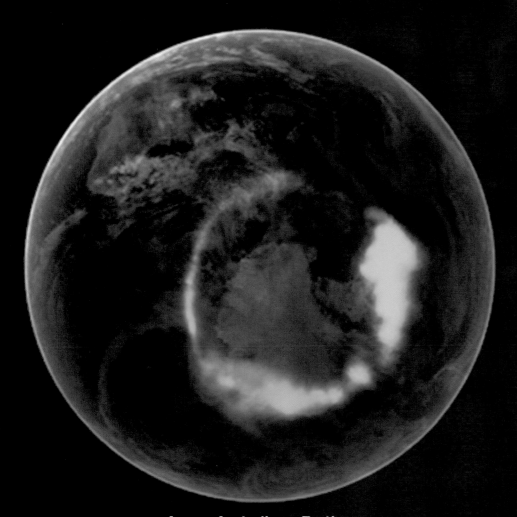

Aurora Australis on Earth

5,140 km

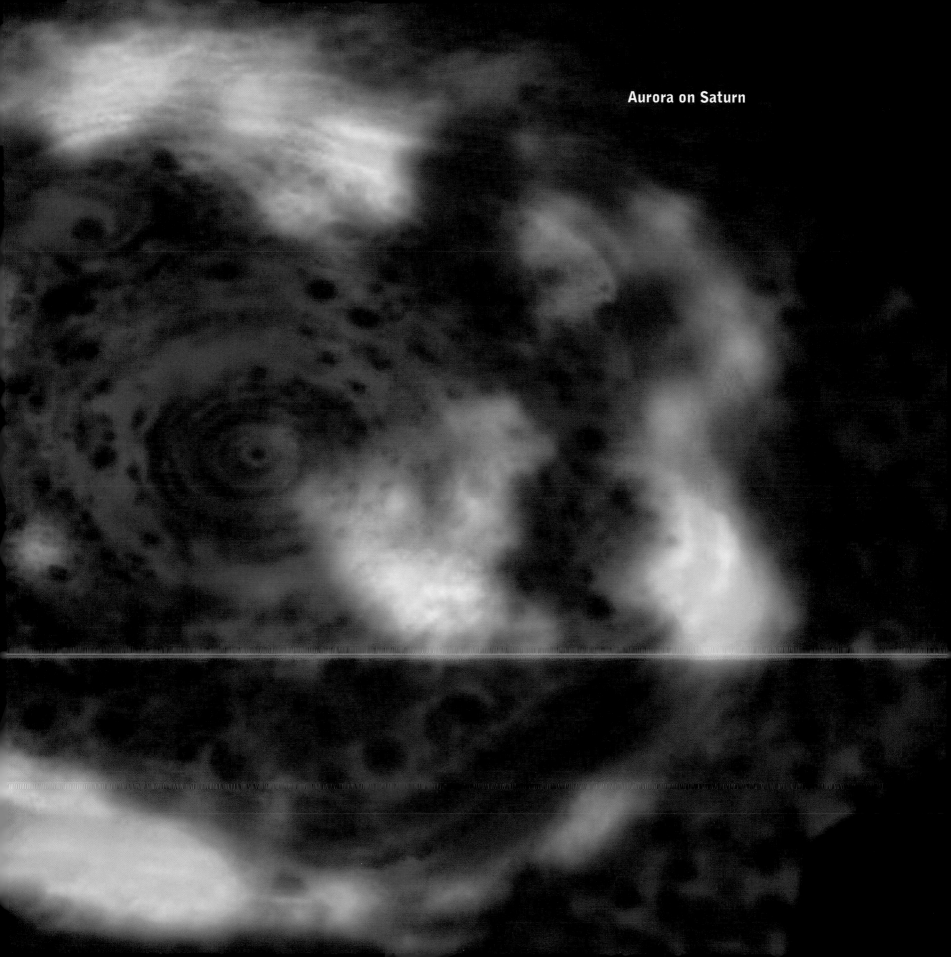

Aurora on Saturn

OTHER SOLAR SYSTEM OBJECTS

IN THIS SECTION, we show all the remaining known objects in the solar system having diameters greater than 500 kilometers (300 miles), along with Earth for comparison. These include moons, Kuiper belt objects, and asteroids, which are all displayed at the same scale we used for the terrestrial planets: 1:100,000,000.

FYI There is more water in the ice-covered ocean of Europa than in all the oceans of Earth.

Saturn has six moons larger than 500 kilometers in diameter. The rings are composed of icy particles ranging from microscopic size to about the size of houses. The rings are less than one kilometer thick. The icy particles making up the rings are tiny moonlets. Saturn's largest moon, Titan, has a substantial atmosphere, and its surface is usually obscured by hydrocarbon smog, as in this picture. Just as Earth has a weather cycle of water (liquid and vapor) in a mostly oxygen-nitrogen atmosphere, frigid Titan (minus 179°C, -290°F) has a weather cycle of methane (liquid and vapor) in a mostly nitrogen atmosphere. In the group portrait on the following two pages we show Titan viewed at infrared wavelengths allowing us to penetrate the smog to see surface features. Titan has lakes of methane near its north pole.

Jupiter's four large moons are Io (3,643 kilometers, 2,264 miles), Europa (3,120 kilometers, 1,940 miles), Ganymede (5,268 kilometers, 3,273 miles), and Callisto (4,800 kilometers, 3,000 miles). Io orbits closest to Jupiter. As it orbits, the gravitational pull of the other satellites tug at it, pushing it first closer to Jupiter and then farther away. Closer to Jupiter, the tidal forces from Jupiter are stronger—squeezing Io into a slightly elliptical shape; farther away, the tidal forces are less, allowing it to become more spherical. This continual kneading heats the interior of Io, melting rock and producing lava that pushes its way to the surface; therefore, Io is covered with volcanoes.

Europa suffers similar effects, but to a lesser degree. Tidal heating of Europa's interior melts deep ice, giving Europa an 80-kilometer-deep ocean beneath a 10-kilometer layer of ice on the top. Since life thrives near hot volcanic vents in the deep oceans of Earth, astrobiologists have wondered whether Europa might also be a place to look for life.

The icy Kuiper belt objects, denizens of the outer solar system, were discovered recently. This portrait illustrates why Pluto (2,306 kilometers, 1,433 miles) was demoted from its planetary status. Many Kuiper belt objects are of similar size, and Eris is larger (2,600 kilometers, 1,600 miles). Pluto is thus a member of this family of icy bodies. Triton has a retrograde orbit around Neptune and may be a Kuiper belt object that has been captured by Neptune.

The rocky asteroids are smaller, with Ceres (975 kilometers, 606 miles) and Vesta (520 kilometers, 320 miles) topping the list. Ceres has, like Pluto and Eris, earned the special designation dwarf planet—a body orbiting the sun, large enough for gravity to determine its shape, but that does not dominate the mass in its orbital region. Although more than 200,000 asteroids are known, only 4 have diameters larger than 500 kilometers (300 miles).

Titan, Saturn's largest moon at 5,150 kilometers in diameter, appears in the background behind Saturn's rings and Saturn's tiny moon Epimetheus (116 kilometers) in the foreground in the Cassini spacecraft picture opposite.

Earth and Moon

for comparison

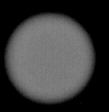

Kuiper Belt Objects

| Eris | Pluto | Haumea | Makemake | Sedna | Charon | OR10 | Orcus |

| T066 | Quaoar | AW197 | MS4 | TX300 | UX25 | Ixion | TL66 | Varuna |

5,080 km

Asteroids

Ceres

Pallas

Vesta

Hygiea

Moons

Neptune

Triton

Uranus

Titania

Oberon

Ariel

Umbriel

Saturn

Enceladus

Tethys

Dione

Rhea

Lapetus

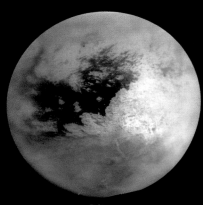
Titan

Jupiter

Europa

Io

Callisto

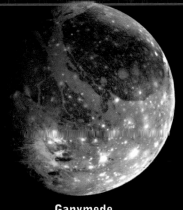
Ganymede

Note: This comparison contains all remaining known objects in the solar system that are larger than 500 kilometers in diameter.

LAKES AND VOLCANOES COMPARED

Lake Michigan

Methane Lakes on Titan

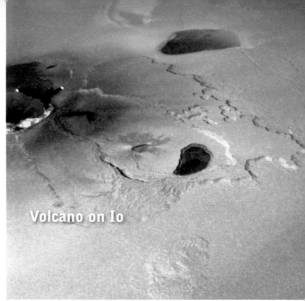

Volcano on Io

Hawaiian Islands

153 km

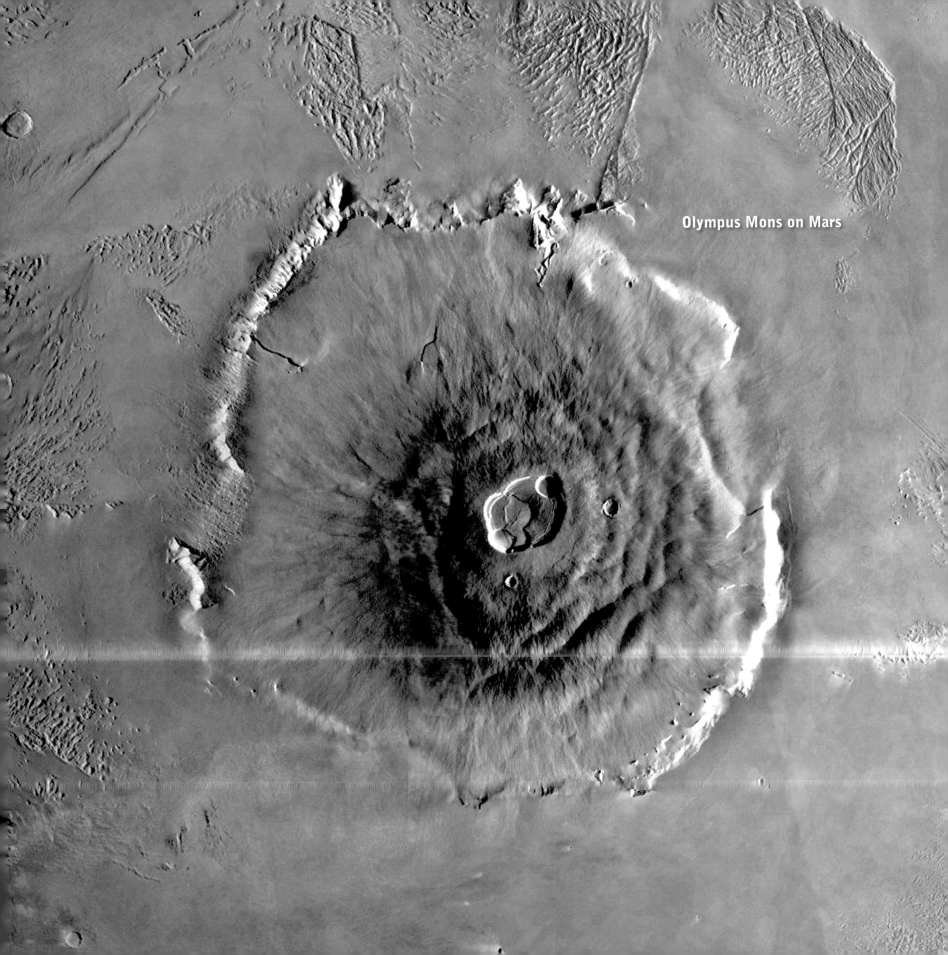

Olympus Mons on Mars

RIVER DELTAS COMPARED

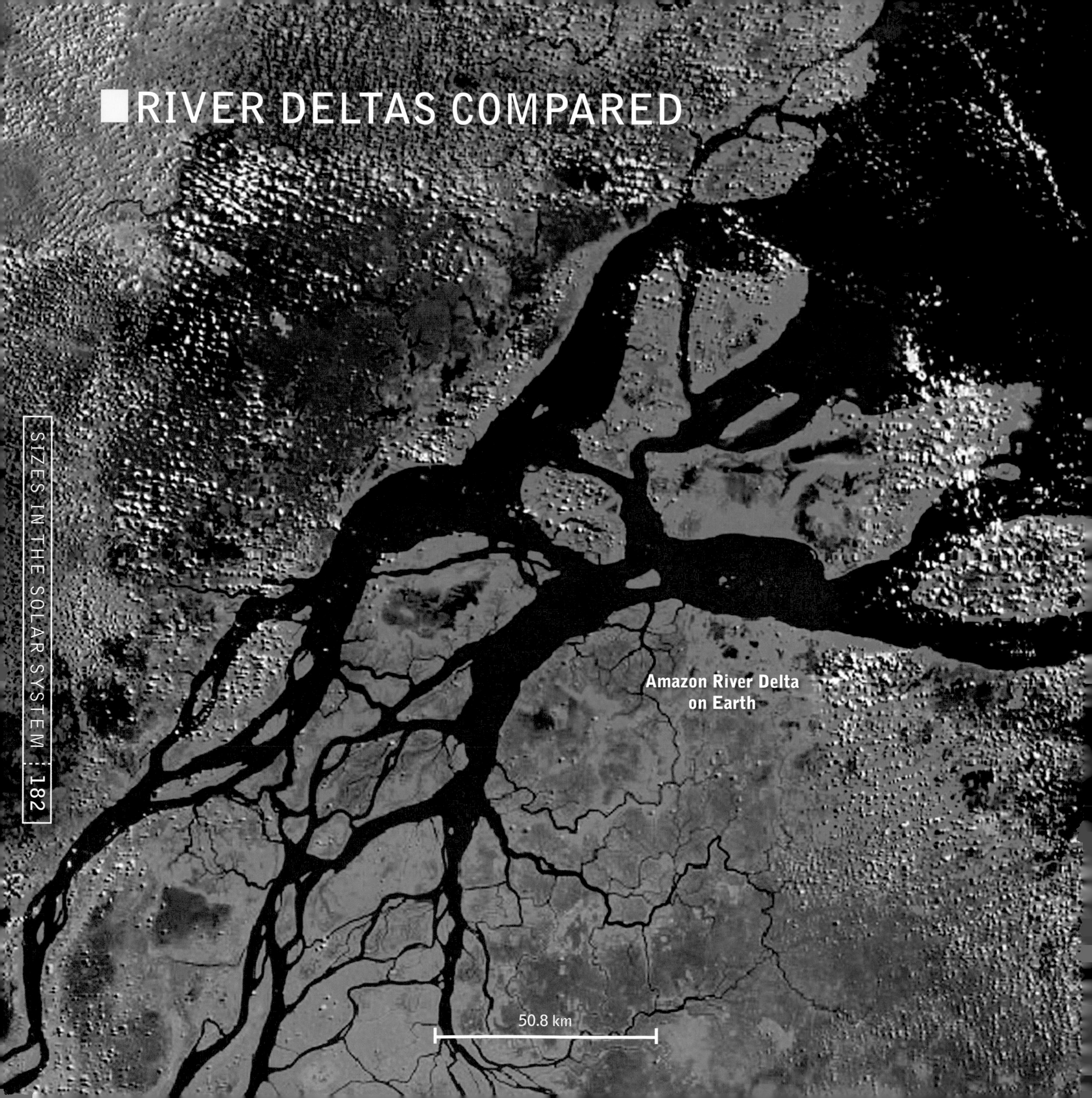
Amazon River Delta
on Earth
50.8 km

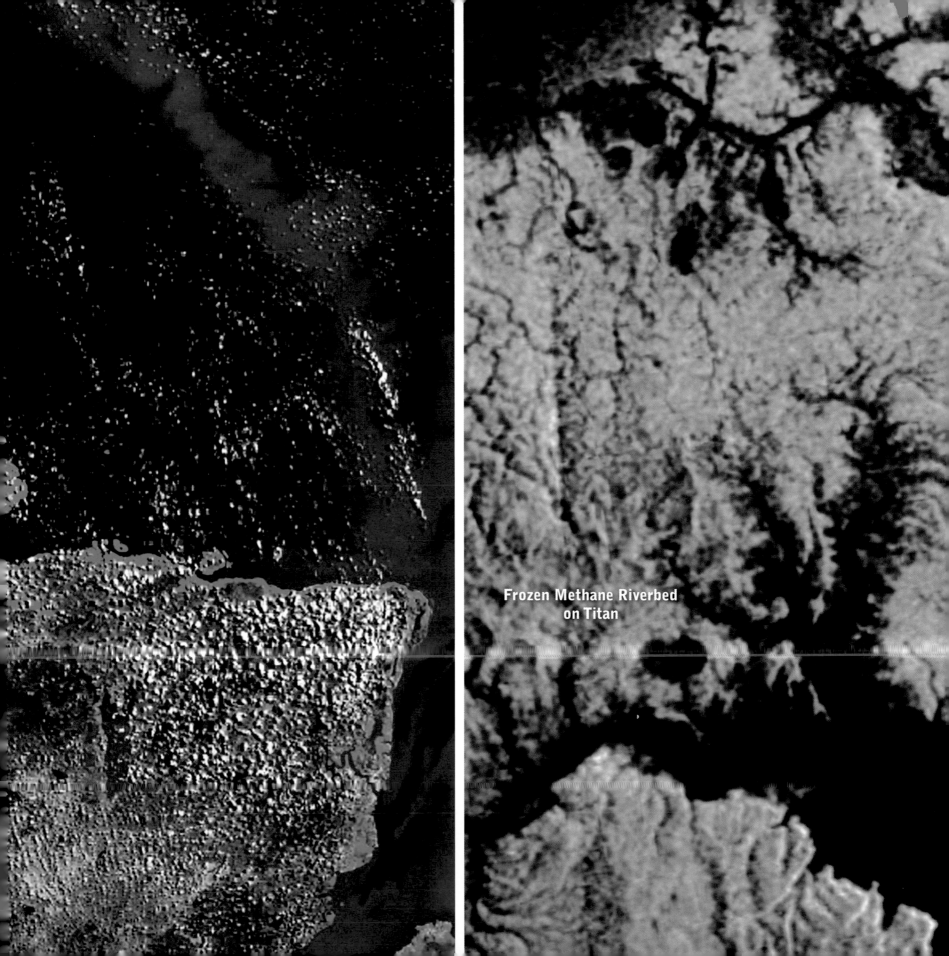

Frozen Methane Riverbed
on Titan

TO INFINITY

CHAPTER 6

AND BEYOND

TEN GIANT STEPS

WE WILL SHOW the sizes of different objects in the universe through a series of scaled pictures. We can get an idea of sizes by looking at scale models or scaled pictures. Doll furniture is usually shown at 1:12 scale, a scale where 1 inch = 1 foot. Model cars are often made at 1:24 scale. Model rockets such as the Saturn V are shown at 1:144 scale, and battleships at 1:720 scale. Maps often appear at a scale of 1:1,000,000, where 1 inch = 16 miles. These scale models and maps can bring large things down to size.

FYI Scientists express very large (or very small) numbers as powers of the base number 10. The Earth's circumference is approximately $40,000,000$ m. In scientific notation, this figure can be shortened to 4×10^7 m.

Actual Size	1:1	Buzz Aldrin's Footprint
1 / Thousand	$1:10^3$	Asteroid
1 / Million	$1:10^6$	Moons
1 / Billion	$1:10^9$	Planets
1 / Trillion	$1:10^{12}$	Stars
1 / Quadrillion	$1:10^{15}$	Solar System
1 / Quintillion	$1:10^{18}$	Globular Cluster M13
1 / Sextillion	$1:10^{21}$	Galaxy M31
1 / Septillion	$1:10^{24}$	Galaxy Clusters
1 / Octillion	$1:10^{27}$	Visible Universe

We will start our journey with a picture shown at actual size, a picture of Buzz Aldrin's footprint on the moon. This picture is 20 inches by 10 inches, big enough to show Aldrin's boot print. The next picture is shown at a scale of 1:1,000 and is labeled 1:1 thousand. It shows a small asteroid together with an astronaut, the space shuttle, the Hubble Space Telescope, and the International Space Station. In the picture is a small, white rectangle 0.02 inch by 0.01 inch showing the size of the previous picture of Buzz Aldrin's boot. You can barely see it. Looking at that tiny, white rectangle showing the size of the previous page gives an idea of just how large a jump that factor of 1,000 is, on what grander scale we are looking when we examine the asteroid. A sequence of pictures, each at 1,000 times smaller scale than the one before, will show how big things are in the universe. Explanations of the objects shown will be found on the overleaf pages.

The Butterfly Nebula, a dying star shedding gas. The ring of dust at the narrow waist, thrown off by the star earlier, now forces newly ejected gas out the top and bottom, creating a nebula that is more than two light-years in diameter.

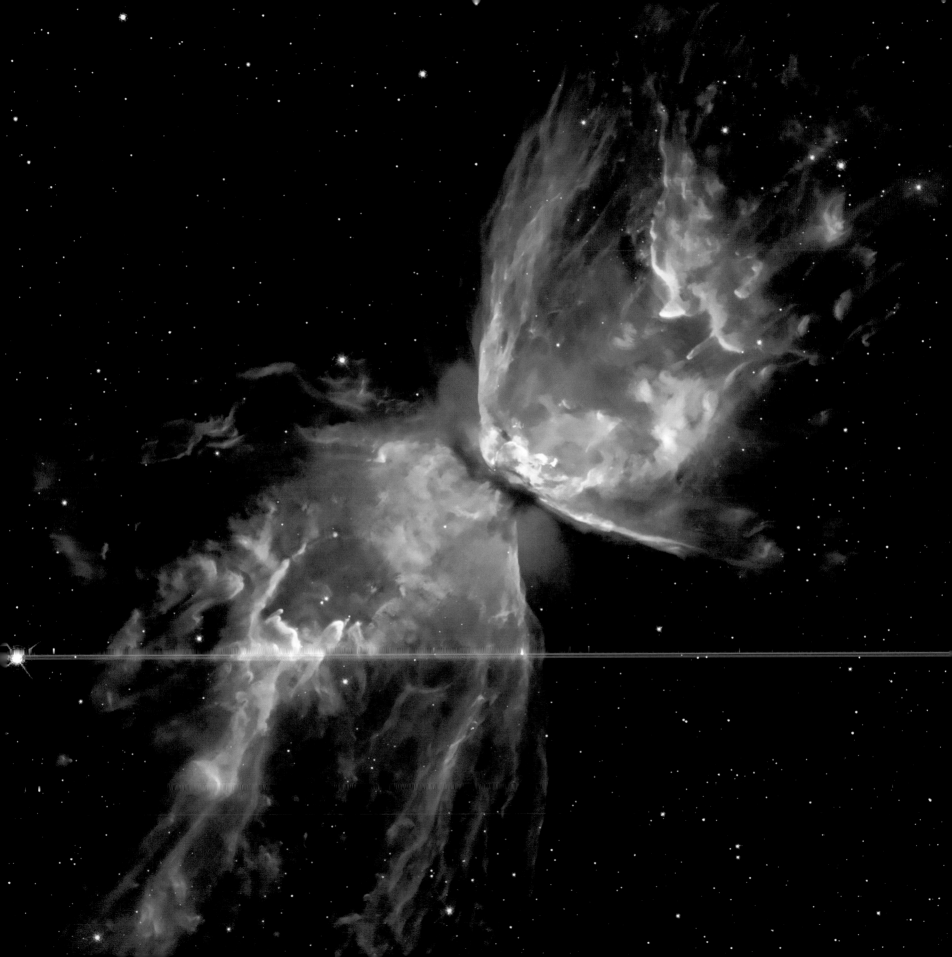

TO SHOW HOW BIG THINGS are in the universe, we will need some really big numbers. In fact, any number that is especially large is often called astronomical, as in "The size of the federal budget this year is astronomical." Fortunately, the decimal system of numbers we use is good at displaying large numbers. Each power of 10 is ten times larger than the one before: 1, 10, 100, 1,000. A thousand is a large number. But

$$1 \text{ million} = 1,000,000$$

is a thousand times larger. And

$$1 \text{ billion} = 1,000,000,000$$

is a thousand times larger still. We can write it as a 1 with 9 zeros after it or we can write it as 10^9 (ten raised to the power 9).

We use the word billion all the time. A billionaire is someone who has a billion dollars. Notice the Latin prefix bi in billion. Bi means two. A bicycle has two wheels. Billion means two factors of a thousand (i.e., 1,000 x 1,000) greater than a thousand. Tri means three, as in triangle. So a trillion is three factors of a thousand larger than a thousand:

$$1 \text{ trillion} = 1,000,000,000,000 = 10^{12}$$

The annual U.S. federal budget is measured in trillions of dollars. Every time we add three more zeros, we get a number a thousand times larger.

$$1 \text{ quadrillion} = 1,000,000,000,000,000 = 10^{15}$$

Quad means four factors of a thousand larger than a thousand. Quintuplets are five babies, so next is

$$1 \text{ quintillion} = 1,000,000,000,000,000,000 = 10^{18}$$

For the next set of numbers, remember that a sextet is a group of six singers, a septuagenarian is someone in their seventh decade of life, and of course an octopus has eight legs. So:

$$1 \text{ sextillion} = 10^{21}$$
$$1 \text{ septillion} = 10^{24}$$
$$1 \text{ octillion} = 10^{27}$$

These are the numbers we will need as we explore the sizes of objects in the universe. Yet mathematicians know still bigger numbers:

$$A \text{ googol} = 10^{100}$$

It is a 1 with a 100 zeros behind it. The search engine Google describes its name as a play on the word *googol*.

$$A \text{ googolplex} = 10^{googol}$$

It is 10 raised to the power of a googol, or a 1 followed by a googol of zeros. The largest number is infinity. But mathematician Georg Cantor distinguished between infinities: The first he called

$$\text{aleph}_0 = \aleph_0 = \text{ the number of counting numbers}$$

1, 2, 3, 4, 5, . . . continue forever—how many numbers are there? There are $\aleph_0$ of them, a countably infinite number of them. But there are still bigger numbers that are uncountable: Ten raised to the power of aleph$_0$— $10^{\aleph_0}$—is the number of points on a line. The number of geometric figures is a bigger number still.

1 million
1,000,000

1 billion
1,000,000,000

1 trillion
1,000,000,000,000

1 quadrillion
1,000,000,000,000,000

1 quintillion
1,000,000,000,000,000,000

1 sextillion
1,000,000,000,000,000,000,000

1 septillion
1,000,000,000,000,000,000,000,000

1 octillion
1,000,000,000,000,000,000,000,000,000

1 googol

10,000,000,000,000,000,
000,000,000,000,000,
000,000,000,000,000,
000,000,000,000,
000,000,000,000,
000,000,000,000,
000,000,000,
000,000,
000

Step One

EVERY JOURNEY BEGINS with a single step. So we shall begin our journey through the scales of the universe with a single step—made by Buzz Aldrin on the moon. Neil Armstrong and Buzz Aldrin were the first men to set foot on the moon. They took off from the Kennedy Space Center in Florida on July 16, 1969. A million people, including one of us (Gott), went to see them off.

The night before, their Saturn V rocket, 110 meters (360 feet) tall, stood on the launch pad illuminated by giant searchlights. The crowd extended up and down nearby Route 1, with cars parked about three or four deep on both sides of the highway. It had the atmosphere of a state fair, with vendors walking up and down the highway all night selling moon cookies and souvenirs. The next morning at 9:32 all was in readiness. As John Barbour of the Associated Press wrote in his book *Footprints on the Moon:* "Men were going to land on the moon. Men were going to walk on the moon. Impossible! The incredibility of it! . . . It wouldn't work. Couldn't work. Shouldn't work. Yet, bit by bit, the realization grew. It would be tried."

When the countdown reached about seven seconds before launch, the five giant engines of the Saturn V ignited, sending plumes of gray smoke out to each side of the rocket as large as itself. Then, as the count reached zero, it was released and ever so slowly began to rise. Faster and faster it climbed, like a magical sword trailing a long plume of flame. The crowd was struck silent by the sight. When the rocket finally disappeared from view behind some high cirrus clouds, people began cheering wildly. Men were going to walk on the moon!

Four days later, on July 20, 1969, Neil Armstrong stepped out onto the lunar surface. Buzz followed shortly after. Neil took a good picture of Buzz's footprint but failed to get one of his own footprint. This image records the high-water mark of human exploration. The next day the *New York Times* printed a banner headline: "MEN WALK ON MOON" in a typeface 1.5 times larger than any they had ever used before.

FYI It's no accident that the "actual size" picture to follow (Aldrin's footprint) is human size. This book is human size by definition—so you can hold it.

The footprint is shown actual size. Human beings are about a billion times bigger than an atom, and as we shall see, about a billion times smaller than a planet. This is no accident. Martin Rees, Bernard Carr, and William Press have calculated that given the strength of the force of gravity and the electromagnetic force, this is about the largest a creature could be—otherwise it would break if it fell over while walking around on its planet. (If gravity were a trillion times stronger, you would break more easily if you took a fall on your planet and would have to be smaller. Your planet would be smaller too; creatures walking on their planets would have to be only a million times larger than atoms, and planets would be only a million times larger than the creatures.) It is fortunate that gravity is weak, for it allows large, complex creatures like us (roughly a billion atoms across) to walk on the surface of their large planet without fear of falling. Thus, a footprint anchors the scale of sizes in the universe, from atoms to planets and beyond.

Buzz Aldrin on the moon, photographed by Neil Armstrong. You can see Armstrong reflected in the visor.

Actual Size

Buzz Aldrin's Footprint
on the Moon

2 in

Step Two

THE PICTURE on the next spread shows a view a thousand times larger than the previous picture. It is made at a scale of 1:1,000 or $1/10^3$. Here, one millimeter represents one meter. The small, white rectangle (near the scale at the top of the picture), which is 0.02 inch by 0.01 inch, shows the size of the previous picture of Buzz Aldrin's footprint on the moon at the current scale. It is tiny but large enough to see.

A number of interesting objects are shown. First is astronaut Bruce McCandless out on a space walk. With a magnifying glass, you can just make out his legs and backpack. Interestingly, years earlier McCandless had been the astronaut at mission control in Houston talking to Neil Armstrong by radio when he took his historic first step on the moon. Next to him is the space shuttle. You can see its open cargo-bay door, its wings, and the front windows for the pilots. You can imagine placing astronaut McCandless inside.

FYI The Hubble Space Telescope is the brainchild of Lyman Spitzer, who conceived of it in 1946 and waged a long battle to get it funded and launched.

In the lower right corner is the International Space Station. Although everyone has heard of it, not very many people would be able to say exactly what it looks like, and this is a particularly good high-resolution picture. Additions continue to be made to the station. The central white column is composed of connected crew modules. They were assembled a piece at a time to make one long passage the astronauts can float through from one end to the other. The large orange rectangles are solar cell arrays

that provide power to the station. Shuttles and Soyuz space vehicles dock with the station to load and unload astronauts and supplies.

Until recently one could sometimes see the shuttle and the space station flying in formation at twilight after a docking mission. The station looks like a bright yellow star (because of the orange solar cells), while the shuttle appears as a dimmer white star.

The Hubble Space Telescope has a 2.4-meter-diameter main mirror at the back (2.4 millimeters wide on this scale). The flap at the front-left open end of the telescope is a sun shield. It was taken up into orbit by the shuttle. You can see that it will just fit snugly into the shuttle cargo bay.

Asteroid Itokawa is the smallest asteroid for which we have a good picture. It is about half a kilometer long (half a meter at this scale). We are looking at it end-on in this picture. It is a rubble pile, debris held together by gravity. You can see boulders of various sizes enmeshed in dirt. Its surface is reminiscent of the lunar soil in the previous picture, but at a thousand times larger scale. This asteroid is not massive enough for gravity to force it into a spherical shape. If you stood on asteroid Itokawa and pushed off hard with your legs you would be able to achieve escape velocity—you would sail away and never fall back.

The International Space Station in low Earth orbit. Here it is at a more advanced stage of completion than in its picture on the following pages.

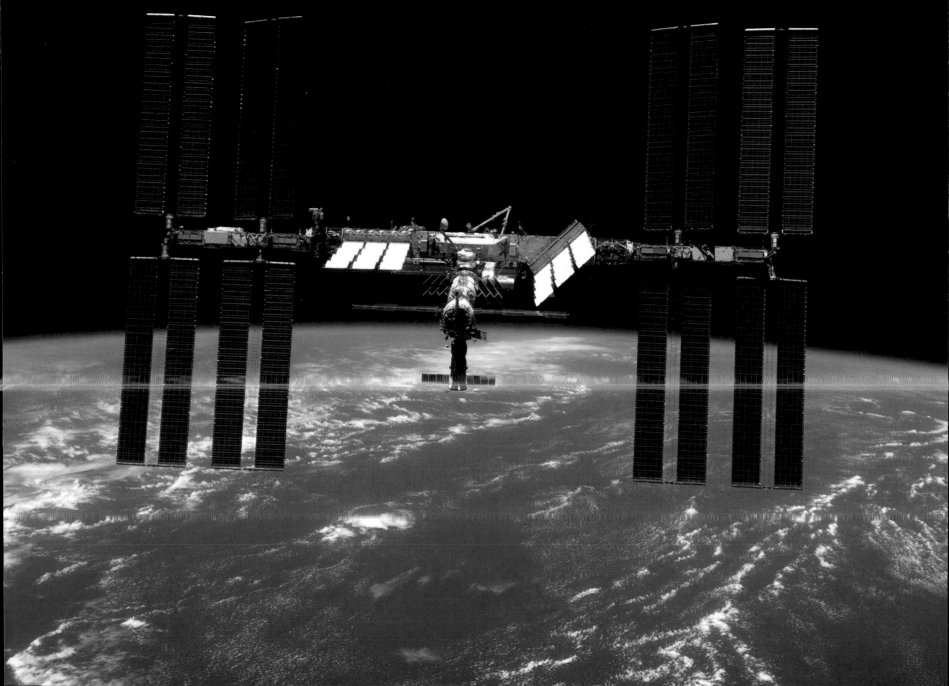

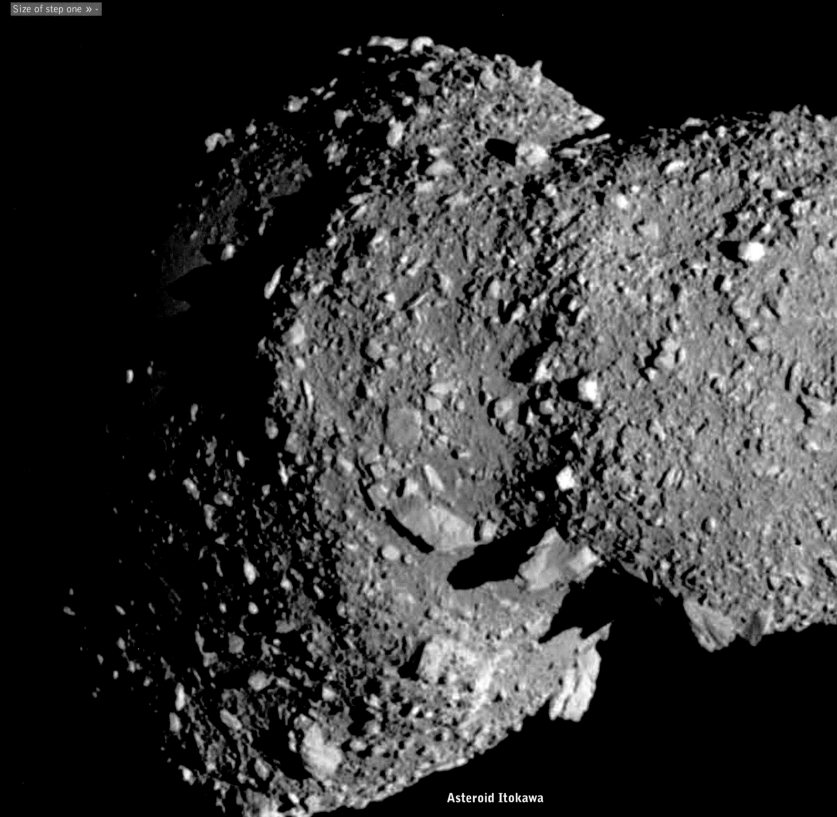

Asteroid Itokawa

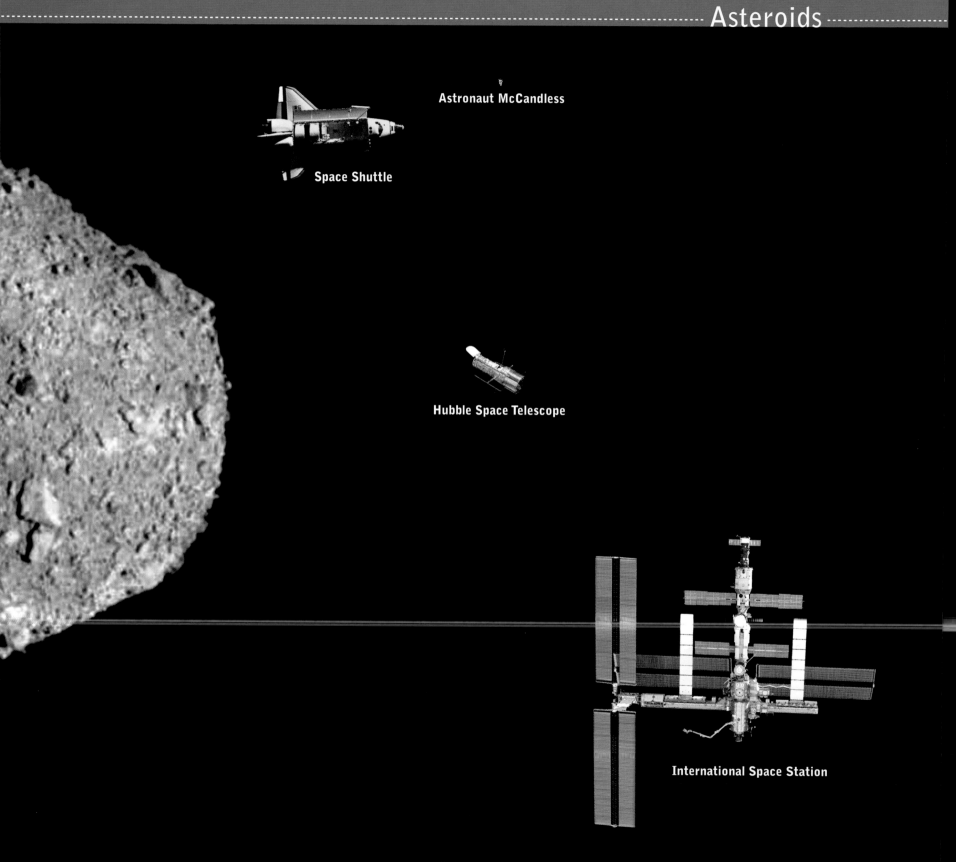

Space Shuttle

Astronaut McCandless

Hubble Space Telescope

International Space Station

50.8 m

Step Three

ON THIS SCALE, 1:1 million, Earth's diameter would be 12.8 meters. A number of interesting objects in space are shown. The asteroid Gaspra seems to be a single rock instead of a rubble pile. Ida is the first asteroid discovered to have a satellite—tiny Dactyl, only 1.4 kilometers in diameter. Ida rotates once every 4 hours, 38 minutes, while Dactyl circles it once every 37 hours. An asteroid about 10 kilometers across—1 centimeter wide on this scale—hitting Earth about 65 million years ago is thought to have caused the mass extinction that killed the dinosaurs.

Deimos and Phobos are the moons of Mars discovered by Asaph Hall in 1877. Remarkably, Jonathan Swift had predicted their existence in his satire *Gulliver's Travels* in 1726. In addition to visiting Lilliput—a land whose people stood only six inches tall, Gulliver also visited the flying island of Laputa. Astronomers there had discovered two satellites of Mars with orbital periods of 10 hours and 21.5 hours. Phobos and Deimos have orbital periods of 7 hours, 38 minutes and 30 hours, 18 minutes, respectively—pretty close to Swift's guess. Phobos and Deimos are so small that in 1959 Russian astrophysicist Iosif Shklovsky speculated they might have been artificial satellites launched by a Martian civilization. However, now we can see that they look like asteroids, probably captured from the main asteroid belt. Their spectra, reflectivity, and density show they resemble a family of carbon-rich asteroids found there.

Comet Wild 2 and Halley's comet are dirty snowballs made mostly of water ice. The picture of Halley's comet shows outgassing vapor and expelled dust emerging from the comet's surface. As comets approach the sun, they become heated, and outgassing vapor gives them long visible tails—millions of kilometers long.

In the cold outer reaches of the solar system we find Enceladus, an icy moon of Saturn, large enough that gravity has pulled it into a nearly spherical shape. Below its icy surface, heating from radioactive decay, as well as tidal heating, most likely produces a liquid ocean beneath a layer of ice—as also occurs on Jupiter's moon Europa. Cratering at top right indicates the icy surface is old, but to the left one sees fewer craters, indicating that the ice has melted and has refrozen. Fractures there show recently formed smooth "blue" ice not yet covered by fine-grain ice particles falling from Saturn's E ring. Fractures like this near the south pole, called tiger stripes, are outgassing water vapor at about 0°C (32°F). This geyser-like venting produces ice crystals that fall like snow, coating the whole moon, giving it a near 100 percent reflectivity. It also contributes micron-size particles to form Saturn's faint E ring—whose densest part is centered on Enceladus's orbit.

A neutron star, a quite different object, is formed as a massive star's core collapses—producing a supernova explosion and ejection of the outer layers of the star. It may be thought of as a giant atomic nucleus stabilized by gravity. If the core of the star is more than twice the mass of the sun, the forming neutron star collapses to form a black hole. Cygnus X-1 is a black hole nine times the mass of the sun. If you were to go inside the event horizon (shown by the red circle), you would never escape.

A simulated close-up view by Andrew Hamilton of how an isolated black hole in our neighborhood of the galaxy would look, including the effects of gravitational lensing of background stars.

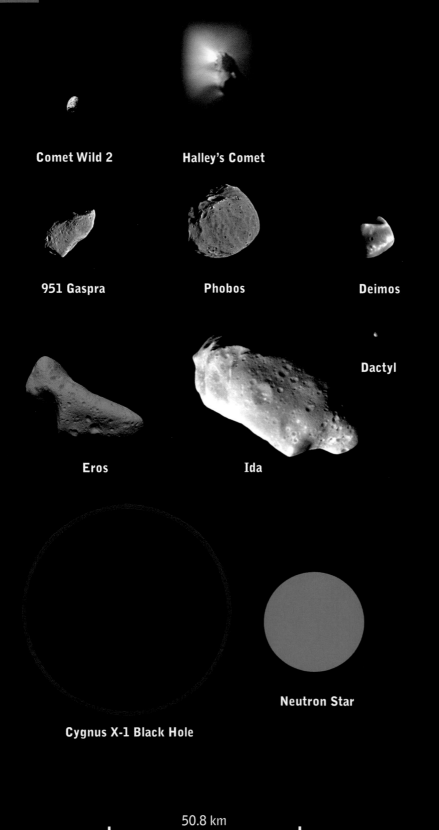

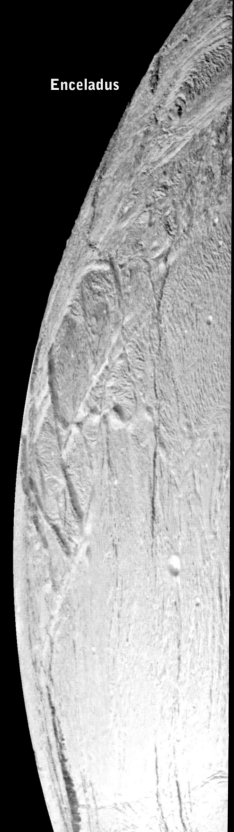

Comet Wild 2

Halley's Comet

Enceladus

951 Gaspra

Phobos

Deimos

Dactyl

Eros

Ida

Cygnus X-1 Black Hole

Neutron Star

50.8 km

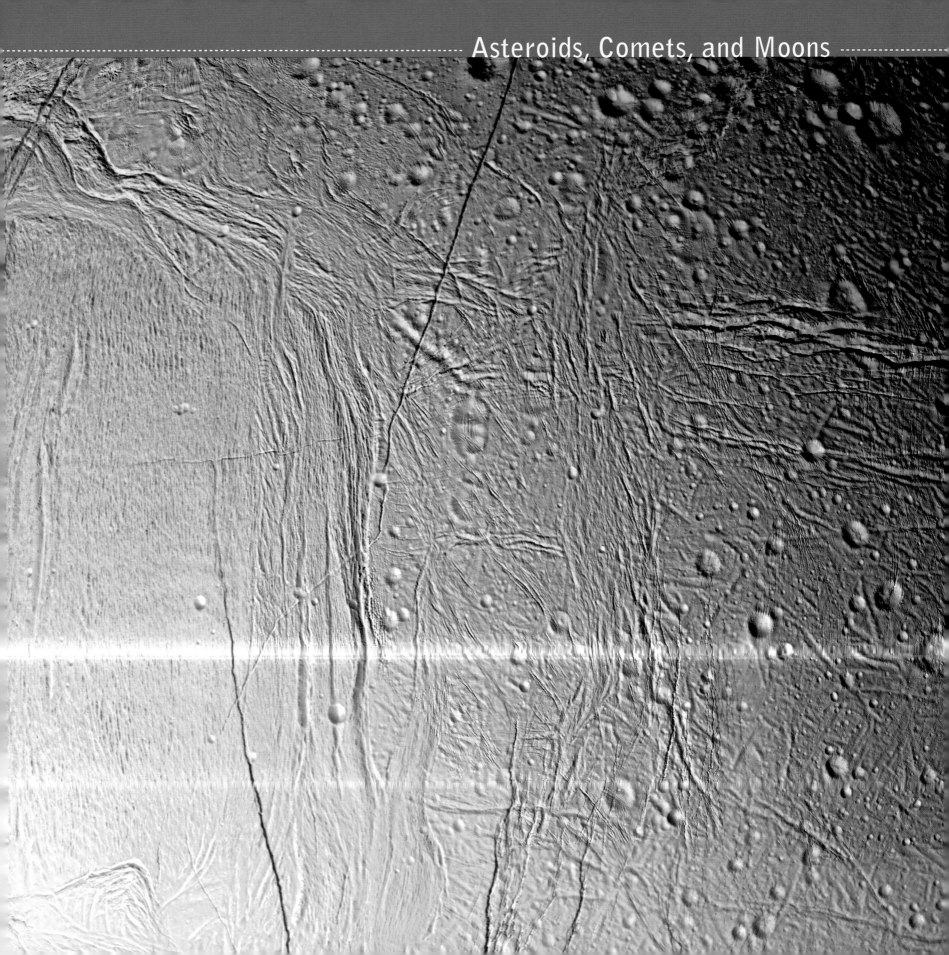

Step Four

HERE WE SHOW all known objects in the solar system larger than 254 kilometers in diameter—objects too large to have been shown completely in the previous picture. At this scale, Earth has a diameter of 12.8 millimeters, or about half an inch.

Peeking in from the right edge is the sun. The sun's diameter on this scale is 1.39 meters. It is so large a million Earths could fit inside. The sun is a ball mostly of hydrogen gas. In its core, hydrogen burns into helium, producing the energy that will allow it to shine for about 10 billion years. You can see sunspots on its surface; these are roughly the size of Earth.

The four rocky terrestrial planets, Mercury, Venus, Earth, and Mars, are shown. So is our moon. Mars's two moons are too small to be shown—we saw them in the previous picture. A swath of rocky asteroids appear, including the largest—Ceres. The four gas giant planets—Jupiter, Saturn, Uranus, and Neptune—are displayed with their large moons. Jupiter's moon Ganymede and Saturn's moon Titan are both larger than Mercury. Saturn's dramatic rings are made of myriad tiny icy moonlets (from microscopic to perhaps the size of houses) orbiting it. The rings of Uranus are also visible. Jupiter and Neptune have rings as well—too faint to be seen here. Jupiter's red spot, a centuries-old storm, is larger than Earth is.

Known icy Kuiper belt objects are shown, the most famous being Pluto and its moon Charon. This picture explains why little Pluto lost its planetary status in 2006, being demoted to dwarf planet. It was the discovery of many other icy bodies like it in the same region, including a larger one, Eris.

We also show the nearest star, after the sun—the red dwarf star Proxima Centauri, as well as the nearby white dwarf star Sirius B. Sirius B was the first white dwarf star studied. It orbits Sirius, the brightest star in the sky. Since Sirius B is very small it is relatively dim. When the sun has used up its hydrogen fuel, after a red giant phase, it will eventually shrink to become a white dwarf. A forming white dwarf more massive than 1.4 solar masses will collapse to form a neutron star or black hole.

FYI At the scale of 1:1 billion, Earth is the size of a marble and Jupiter is about the size of a large grapefruit.

We have gathered these objects together in one place to compare them in size. At this scale the moon is about 38 centimeters from Earth, the sun is about 0.15 kilometers away, Proxima Centauri is about 40,000 kilometers away, and Sirius B is more than 80,000 kilometers away.

The dramatic 3-D image of Saturn and its rings (opposite) is made from two pictures of Saturn taken by the Hubble Telescope at different times. Particles orbiting at the inner edge of the Cassini division (the dark, narrow gap in the rings) are in a two-to-one resonance with Saturn's moon Mimas, circling Saturn twice every time Mimas circles once. (View cross-eyed for 3-D. See pages 98-99 for instructions on cross-eyed viewing.)

Periodic perturbations by Mimas cause particles orbiting just outside this radius to spiral outward, partially clearing the Cassini division of particles and making it appear dark. In this picture we can see clearly the A ring (outside the Cassini division), the white B ring (inside the Cassini division), and inside that, the translucent C, or Crepe, ring which can be seen against Saturn's disk.

The standard theory of Saturn's rings has been that they were formed when an errant moon wandered too close to Saturn (inside what is called the Roche limit) and was ripped apart by tidal forces. But Saturn has no wandering moons. All its regular moons out to Titan are composed primarily of water ice just like the rings and are in nice circular orbits.

We (Gott and Vanderbei) have argued, along with our colleague Edward Belbruno, that Saturn's rings as seen today are the remnant of an originally much larger ring system. Outside the Roche limit, icy moons were able to form from the icy ring particles, but inside the Roche limit, tidal forces prevented the formation of large moons,

leaving the rings we see today. In this model, the rings would be old. Recently, the Cassini spacecraft has found new evidence suggesting that the rings are indeed old. The rings are relatively clean of contamination by meteoric dust, which suggested a more recent origin. The Cassini spacecraft, however, found the rings to be substantial enough to dilute the dust, with particle collisions continually breaking apart particles, uncovering new, fresh ice surfaces relatively clean of dust. Thus, the rings can be old.

Two views of Saturn taken by the Hubble Space Telescope at different times. They give slightly different viewing angles and can be combined to make a dramatic 3-D image.

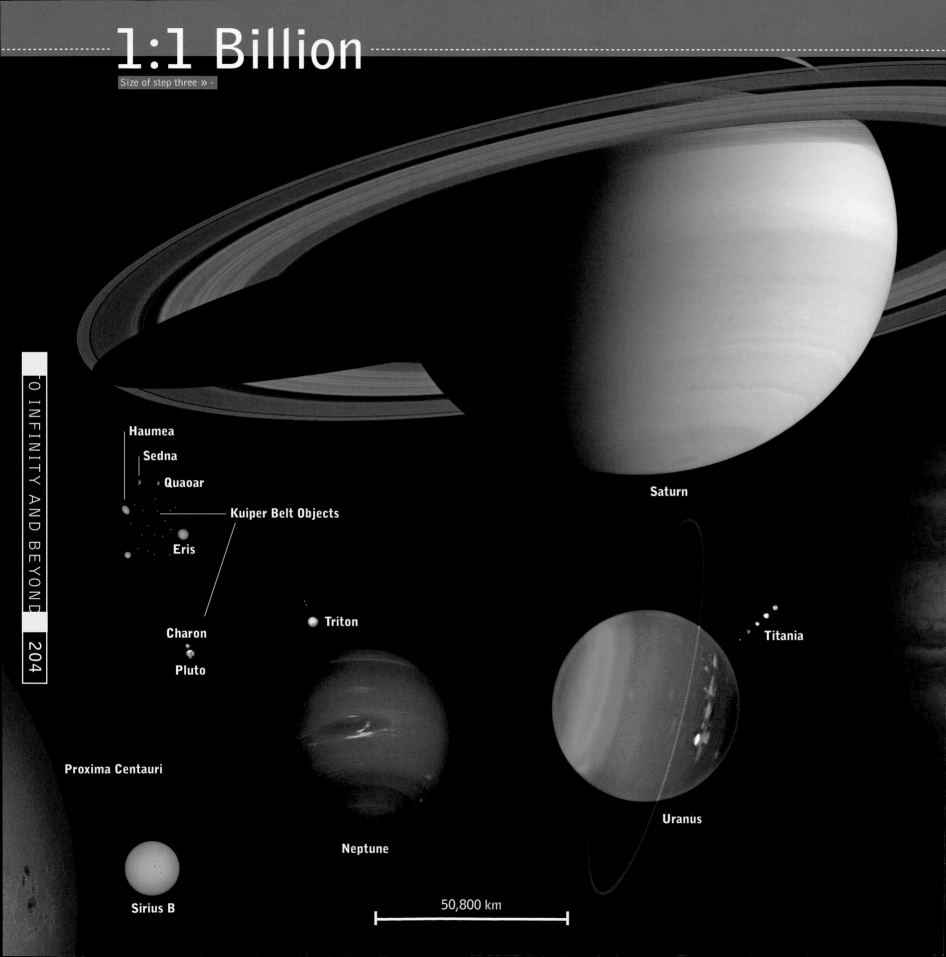

Haumea

Sedna

Quaoar

Kuiper Belt Objects

Eris

Saturn

Charon

Triton

Titania

Pluto

Proxima Centauri

Uranus

Neptune

Sirius B

50,800 km

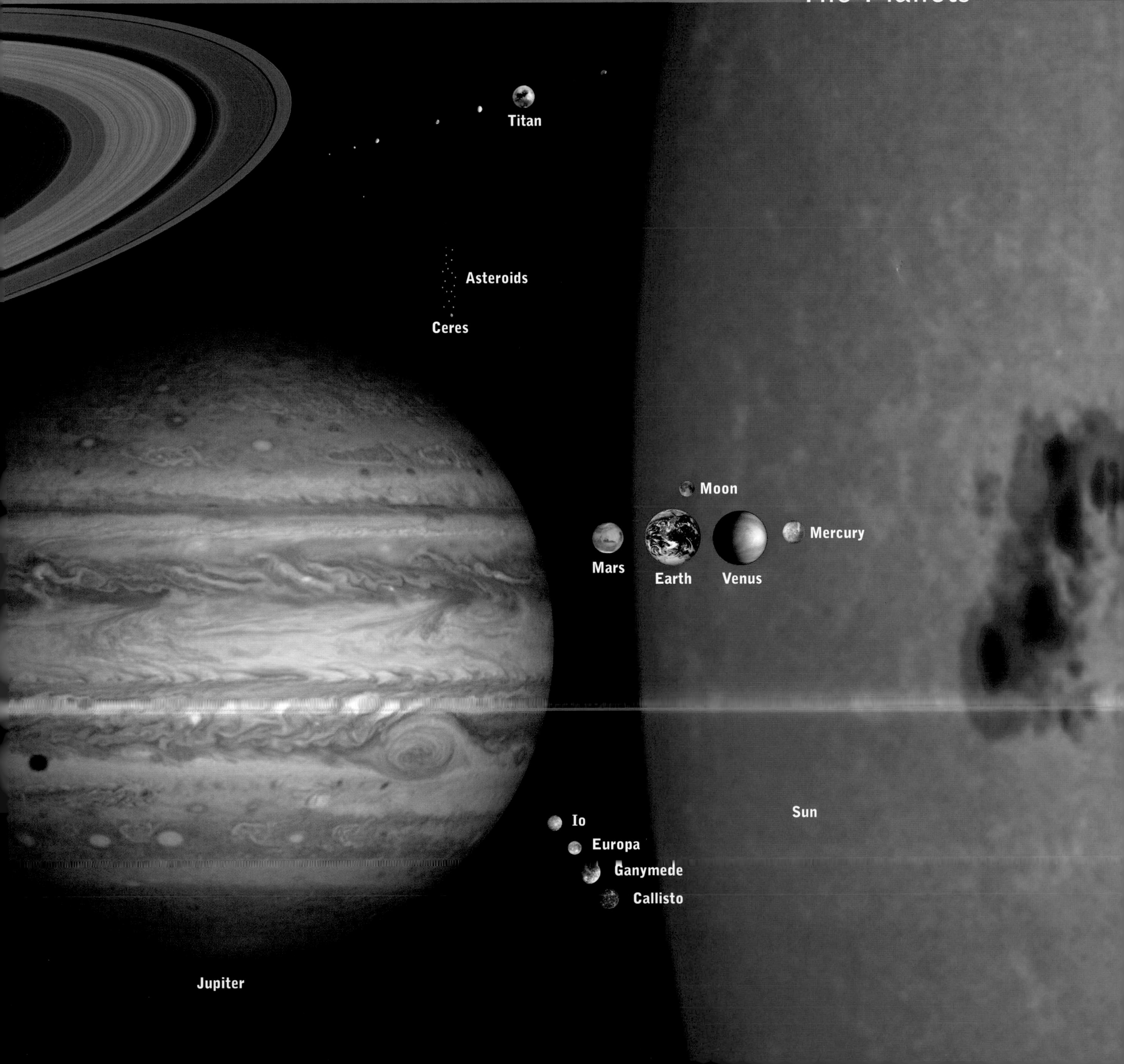

Titan

Asteroids

Ceres

Moon

Mars

Earth

Venus

Mercury

Io

Europa

Ganymede

Callisto

Sun

Jupiter

PLANETS ARE SHOWN to scale in silhouette against their stars as if seen in transit. The sun and its planets, Pluto, and some moons are shown for comparison. We can discover the sizes of extrasolar planets by noting the fraction of their star's light they block if they transit in front of it. Most planets discovered to date are very close to their stars and hence too hot to allow liquid water on their surface. Planet HD 209458b is a hot gas-giant planet like Jupiter. Planet GJ 436b is a hot Neptune-like planet. It's hot because it is so close to its star, even though that star is a cool M-dwarf. CoRoT-7b is the smallest transiting planet discovered so far—its diameter is only 1.7 times greater than Earth's diameter. It is a rocky planet with a temperature of more than 1300K.

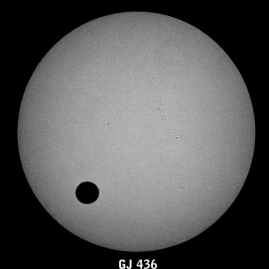

GJ 436

CoRoT-7

508,000 km

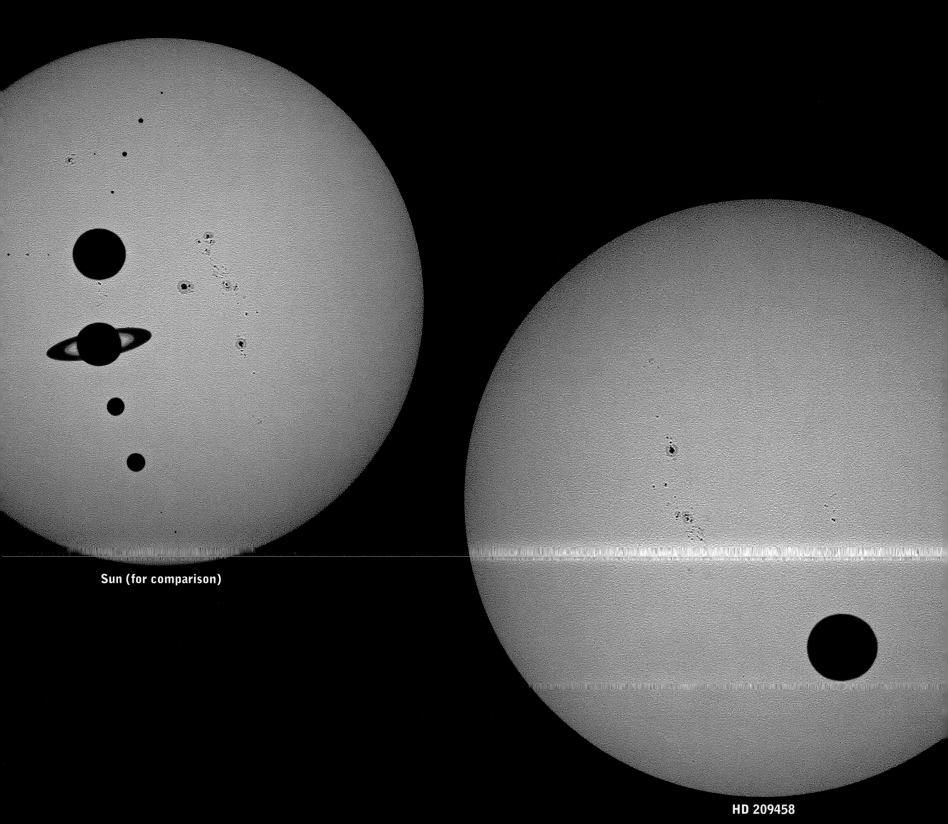

Sun (for comparison)

HD 209458

Step Five

AT THIS SCALE, we see stars. The sun, here with a diameter of 1.39 millimeters, appears with the orbits of Mercury, Venus, Earth, and Mars for comparison. We show all the stars in their true colors as seen in visible light. The sun is white, with a surface temperature of 5800K.

Other stars, burning hydrogen in their cores like the sun, also appear. Proxima Centauri is a red dwarf with a mass of 0.12 M_{sun} (where M_{sun} is a solar mass) and a total luminosity of 0.0014 L_{sun} (where L_{sun} is a solar luminosity). Its surface temperature is 3040K, giving it a red color. Its low luminosity means it is burning its hydrogen fuel slowly, so it will still be shining long after the sun has died. Other main-sequence stars are :

Gliese 581 (0.31 M_{sun}, 0.013 L_{sun}, 3500K),
Tau Ceti (0.81 M_{sun}, 0.59 L_{sun}, 5300K),
HD 209458 (1.1 M_{sun}, 1.6 L_{sun}, 6000K),
Alpha Centauri (1.1 M_{sun}, 1.5 L_{sun}, 5800K),
Sirius (2.0 M_{sun}, 25 L_{sun}, 9900K),
Vega (2.6 M_{sun}, 51 L_{sun}, 9300K), and
Regulus (3.5 M_{sun}, 150 L_{sun}, 10, 300–15,400K).

Sirius, Vega, and Regulus are so hot they are blue—like the blue-hot flame of a blowtorch. Regulus is rotating so rapidly that it is pulled outward at the equator into an elliptical shape. Stars more massive than the sun burn out more quickly than the sun.

For Gliese 581, note the orbit of Gliese 581c, the third of its four planets. This planet is probably just a bit too hot to retain liquid water on its surface. Gliese 581d (somewhat more than seven times Earth's mass) is probably a little cold to have liquid water on its surface *unless* it has a thick CO_2 atmosphere causing a greenhouse effect. HD 209458b

is a planet two-thirds the mass of Jupiter. Its orbit, close to a star more luminous than the sun, gives this planet a furnace-like temperature of about 1000°C (1832°F).

Rigel (17 M_{sun}, 40,000 L_{sun}, 11,000K) is a blue supergiant, and Betelgeuse (18 M_{sun}, 40,000–100,000 L_{sun}, 3500K) is a red supergiant. Both have exhausted the hydrogen fuel in their cores and have moved off the main sequence, burning hydrogen in a shell and burning heavier elements in the core. These two massive stars are expected to eventually explode as supernovae, their cores collapsing to form neutron stars. When the sun exhausts the hydrogen fuel in its core, it will become a red giant as large as Earth's orbit. It will eject some gas and eventually collapse to form a white dwarf.

FYI For main sequence stars (like the sun and those listed at left), the more massive they are, the more luminous and the hotter.

Lurking at the left is the black hole at the galactic center (4.5 million M_{sun}). Its event horizon is shown in red. Comet Hyakutake, with its tail of outgassing molecules and dust, is shown as it appeared during its close approach to Earth in 1996.

Hertzsprung-Russell diagram. Stars are plotted as a function of their temperature (hotter ones to the left) and luminosity (brighter ones at the top). Diagonal lines are lines of constant radius.

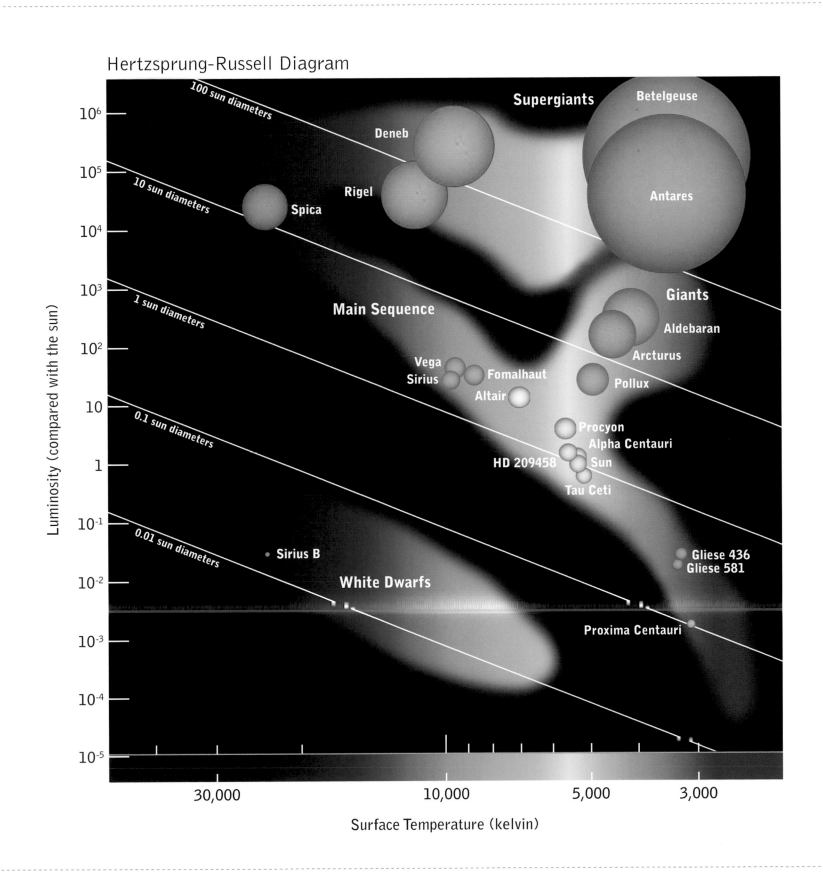

Hertzsprung-Russell Diagram

Black hole at
galactic center

Comet Hyakutake

Rigel

Mars's orbit

Orbit of d

c

b

e

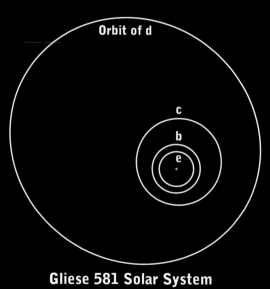

Gliese 581 Solar System

Regulus

Vega

Sirius

Alpha
Centauri

Tau Ceti

Proxima
Centauri

b's orbit

HD 209458

50,800,000 km

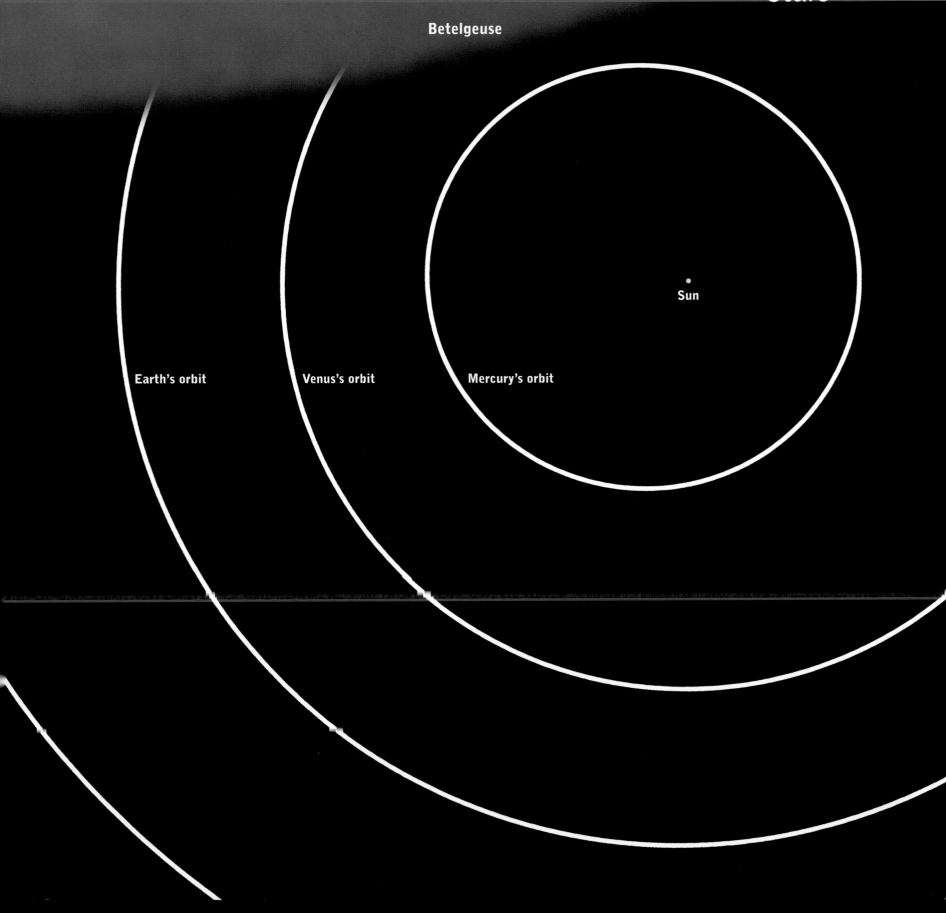

Stars

Betelgeuse

Sun

Earth's orbit

Venus's orbit

Mercury's orbit

Step Six

NOW THAT WE HAVE jumped another factor of a thousand in scale, we can see all of Betelgeuse. It is one of the largest stars—yet it is small at this scale. In the previous step, we saw the orbits of the inner planets out to Mars. Now the whole solar system comes into view. The orbits of Jupiter, Saturn, Uranus, and Neptune are shown circling the sun. The highly elliptical orbit of Halley's comet is shown in white. It orbits the sun once every 76 years. Beyond Neptune's nearly circular orbit are the Kuiper belt objects Pluto, Eris, and Sedna, which travel along more eccentric orbits. Sedna is an icy body that takes about 12,000 years to complete each orbit around the sun. One "Sedna-year" ago, humans had not yet begun to do agriculture.

FYI The diameter of a black hole is proportional to its mass. M87 is the most massive black hole we have found so far.

To the left, the Vega dust disk is displayed. The star Vega itself is tiny—we saw it in the previous picture. But Vega is surrounded by a disk of dust as large as our whole solar system—as you can see. We are looking at this disk nearly pole-on, so it appears round. The dust grains circling Vega are thought to be small—each about 0.05 millimeter or less in diameter. They are likely produced by collisions between larger bodies (like our Kuiper belt objects) circling Vega. The total amount of dust is about 1/300 the mass of Earth.

At the lower right, we show the orbit of Fomalhaut b, the first extrasolar planet discovered by direct imaging. It orbits the star Fomalhaut in a nearly circular orbit at a distance of 115 AU. It takes about 870 years to complete one orbit around Fomalhaut.

At the far lower right is the black hole in the galaxy M87. This is the largest black hole we have found so far—it has a mass three billion times the mass of the sun. The red circle shows the circumference of the black hole's event horizon—the point of no return. The escape velocity at the event horizon is equal to the speed of light, and since nothing can go faster than the speed of light, once you cross the event horizon you can never escape. If you free-fall feet first into this black hole from a large distance, from the time you cross the event horizon, it will take you about another five and a half hours to reach the center. At that time, if the black hole was not rotating, you would be ripped apart by tidal forces as your feet, positioned closer to the center, are pulled in more strongly than your head. Fortunately, this period of torture would be brief—only 0.08 seconds from the time the forces began to be uncomfortable until you were completely shredded. A quick end, at least!

But if the black hole is rotating, as we surely expect in this case, you may have some glimmer of hope after all. Tidal forces become infinite only on a ring singularity, a region of infinite space-time curvature, which you can avoid hitting. But you may be killed first, by running into highly blueshifted photons that have gained energy falling into the black hole. They can create a singularity of their own—blocking your way. To learn whether you will be torn apart by this singularity we may have to understand quantum gravity—how gravity behaves on microscopic scales. Singularities may be smeared out and mitigated by quantum effects, and according to some speculative

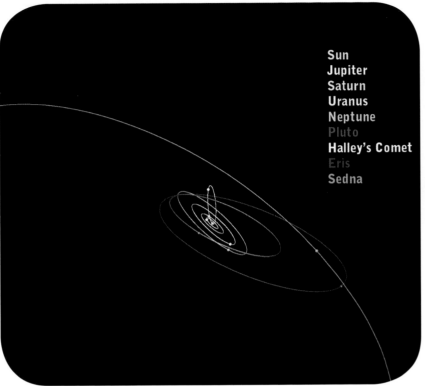

Sun
Jupiter
Saturn
Uranus
Neptune
Pluto
Halley's Comet
Eris
Sedna

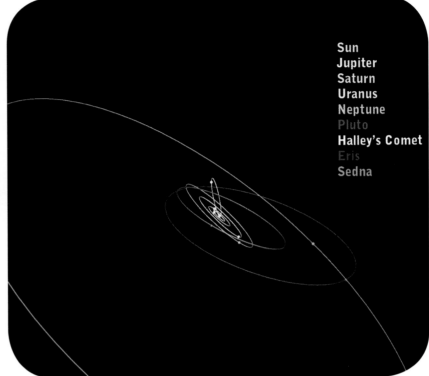

Sun
Jupiter
Saturn
Uranus
Neptune
Pluto
Halley's Comet
Eris
Sedna

ideas, you might even be able to do some time traveling inside the black hole before being squirted out in another universe. But you would never get back to your friends to brag about your adventures—once inside that event horizon, there is no coming back outside. Beware of falling in!

Above we see the outer solar system in 3-D (use cross-eyed viewing technique). The orbits of Jupiter, Saturn, Uranus, and Neptune are all nearly circular and in nearly the same plane—the same plane shared by Earth's orbit. We are observing from just above this plane. The orbits of other bodies are more eccentric and inclined. Halley's comet (in white) has a very eccentric orbit. It returns close to the sun every 76 years. The orbits of three Kuiper belt objects are also shown. Pluto's orbit is tipped at an angle of 17 degrees relative to the plane established by the orbits of the inner planets. Pluto takes more than 248 years to complete each circuit around the sun. Recently

discovered Eris is larger than Pluto, and its orbit is also inclined relative to the plane of the inner planets.

Finally, we can see a small part of the orbit of Sedna sweeping by. Its eccentric orbit carries it much farther from the sun than could be shown in this picture. Sedna was discovered while making its close approach to the sun—otherwise, it would have been too faint for us to have found with current equipment. This 3-D picture captures the complex relationships of these orbits in space, something impossible to appreciate in a 2-D picture. The little spheres show where the bodies were on February 21, 2005.

The solar system in 3-D showing the orbits of the planets, some famous Kuiper belt objects, and Halley's comet.

1:1 Quadrillion

Size of step five » -

Betelgeuse
Vega dust disk
50.8 billion km

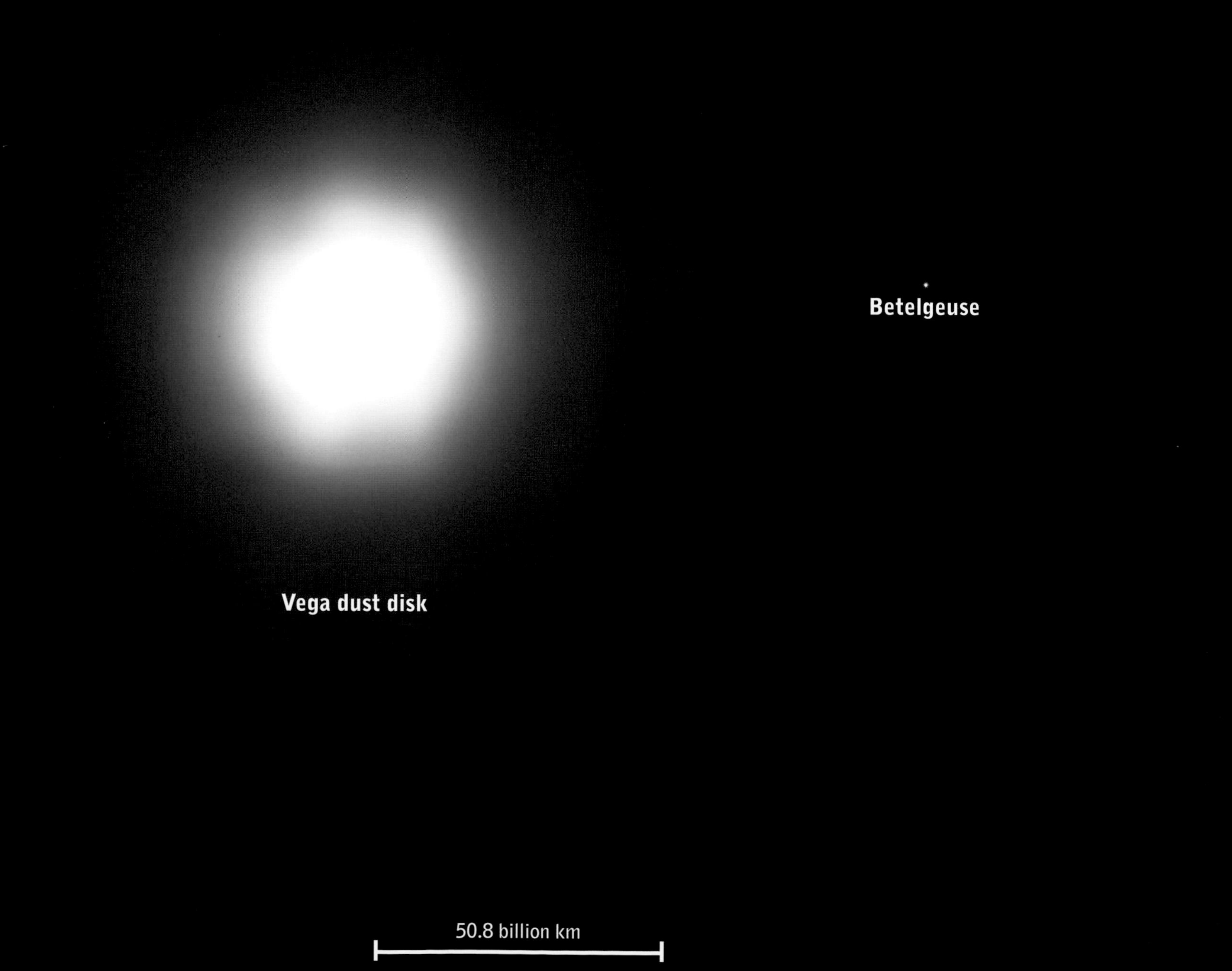

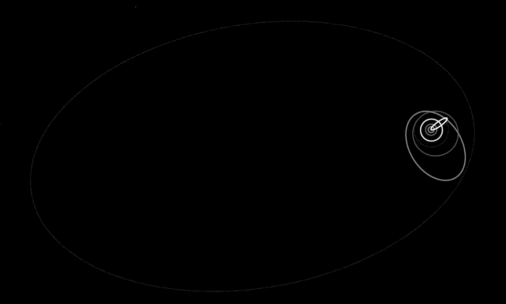

(Orbits, from small to large)

Jupiter
Saturn
Uranus
Neptune
Halley's Comet
Pluto
Eris
Sedna

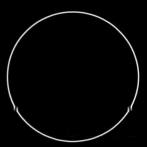

Fomalhaut-b's orbit

Black hole in M87

Step Seven

HERE WE HAVE MOVED up to interstellar distance scales. The sun, Alpha Centauri, and Sirius form a nearly perfect right triangle in space, which is framed in the rectangle to the right. Alpha Centauri is 4.37 light-years from Earth.

Is interstellar travel possible? Sure. The carbon atoms in your body are interstellar travelers from distant supernovae that blew up before the solar system was formed. But it's a long trip. A rocket traveling a thousand times faster than Apollo 11 would still need 115 years to get to Alpha Centauri—piloted by a multigeneration crew.

FYI The distance from the sun to Alpha Centauri is about a billion times longer than the circumference of the Earth.

Could we get there any faster? A rocket accelerating at 1 g (or g-force), so the crew would feel the same weight as they do on Earth, could reach a velocity of 94.8 percent the speed of light after traveling three years and covering half the distance. Decelerating at 1 g for the next three years to slow back down to zero velocity would allow the rocket to reach Alpha Centauri after six years. The crew would age only 3.6 years during the trip; as Einstein showed, rapidly moving clocks tick slowly. Sending a crew of eight, recycling air, food, and water (maintaining life in a closed system), requires a payload of at least 400 tons and, even with maximally efficient matter-antimatter fuel, a rocket of more than 14,800 tons. The energy to produce the antimatter fuel for this one-way trip would be more than 1,300 times the world's current annual energy consumption. Expensive!

Dominating the scene at this scale is the globular cluster M13, containing more than 100,000 stars. The brighter stars are red giants. The Horsehead, Eagle, and Orion Nebulae are dense gas clouds where stars are forming now. Other objects mark the death of stars. The Eskimo Nebula is gas ejected from a dying star. The Ring Nebula is similar—from our line of sight along a polar direction, we see an equatorial ring of ejected gas surrounding the star remaining at the center. The Dumbbell Nebula represents a similar phenomenon seen from an equatorial direction. The Crab Nebula is gas ejected from a supernova observed by Chinese astronomers in 1054. Expanding outward at about 1,500 kilometers per second, the debris has taken almost a thousand years to reach this size. At the center is a pulsar (neutron star) rotating 30 times a second.

Supernova 1987A was a blue supergiant before it blew up. Ultraviolet light from the explosion has now hit rings of gas ejected earlier, causing them to glow. Knowing the angular size of the rings and the observed time delay between the explosion and when the rings were first illuminated, we can estimate the distance to the supernova—168,000 light-years. It lies in the Large Magellanic Cloud, a satellite galaxy of the Milky Way. Some gas from dying stars, traveling through interstellar space, eventually ends up in dense gas clouds—stellar nurseries for the next generation of stars.

Vanderbei's photograph of the Orion Nebula. The nebula is more than 1,300 light-years away.

M13

5.37 ly

SN 1987a

Ring

Eskimo

Dumbbell

Horsehead

. Alpha Centauri

. Sun

Sirius .

Orion

Eagle

Crab

■NEBULAE COMPARED

Rosette

32.4 ly

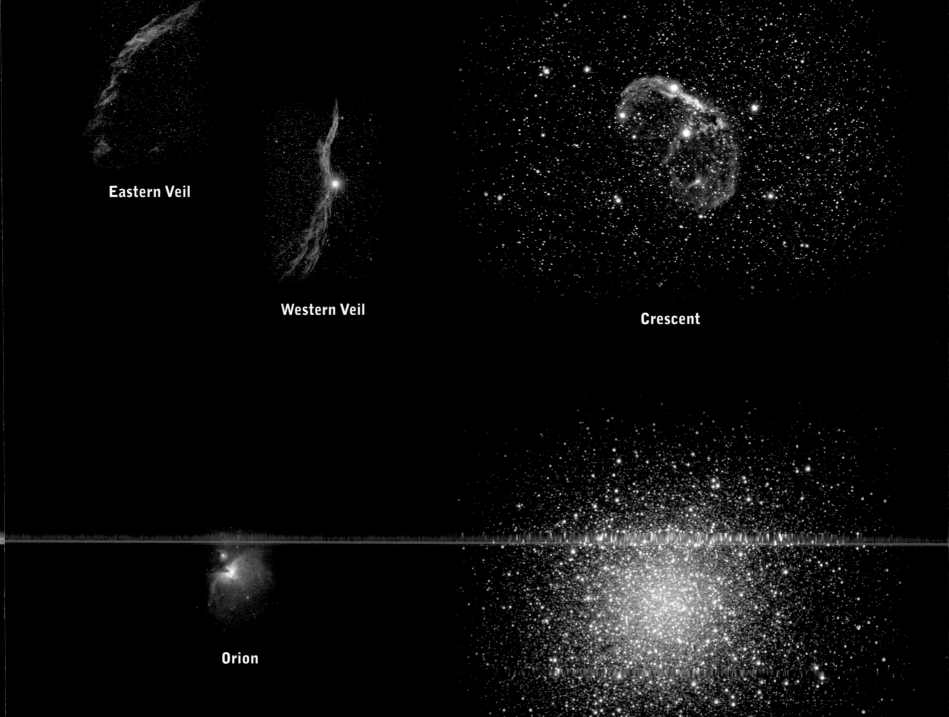

Eastern Veil

Western Veil

Crescent

Orion

M13

Step Eight

AT THIS STAGE in our journey, let us pause to admire the view. The magnificent panorama of the Andromeda galaxy (M31) stretches more than 50,000 light-years from left to right. Ignore the beautiful sprinkling of bright stars that cover the page—they are foreground stars in our own Milky Way, just raindrops on our windshield obscuring the distant view. We want to concentrate on the Andromeda galaxy—but we are seeing only part of it. (You can see its full extent on pages 36-37.) At the scale shown here, it would be more than a meter wide. Andromeda is about 2.5 million light-years away, yet visible with the naked eye on a clear night from a dark location. The photons from it that we see today started on their way just before the time *Homo habilis,* the first species of our genus, began to walk the Earth.

At this scale, the distance between the Milky Way and Andromeda would be 24 meters, the Sloan Great Wall of galaxies would be 13 kilometers long, and the entire visible universe would have a radius of 130 kilometers. The Ring Nebula we saw in the previous picture would be the size of a red blood cell, Earth's orbit would be the size of an oxygen atom, and Earth itself would be the size of the nucleus of an atom of Europium. The moon would be the size of a helium nucleus. The globular cluster M13 that dominated the previous picture would be just a small dot.

The Andromeda galaxy has several hundred globular clusters of its own. We have identified two of the larger ones—G64 and G76. They could be mistaken for foreground stars in the Milky Way. We are looking at the disk of Andromeda from a nearly edge-on perspective. It is a spiral galaxy like our own. We can see one of its two spiral arms—a fishhook of blue stars ending at M32, a satellite galaxy of M31 just off the top edge of the picture. The

spiral arm is bounded by dark dust lanes as it continues to the right. As stars of Andromeda's disk circle the center like cars on an expressway, those on the inner lanes are continually passing those farther out. Gravitational perturbations by the nearby satellite galaxy M32 set up spiral density waves—moving traffic jams of stars. The increased density in these spiral arms causes gas clouds to form, igniting bursts of star formation.

Along the spiral arms are dust clouds and large glowing clouds of hydrogen gas (like the Rosette Nebula)—little red splotches where stars are being born. Bright, massive young blue stars outline the spiral wave during their brief lives. In 1943, Walter Baade discerned two populations of stars in Andromeda. The light of the outer disk is dominated by young, blue, main-sequence stars (Population I), while the light of the central spheroidal bulge is dominated by older red-giant stars (Population II). M110 is a dwarf elliptical galaxy, a satellite of M31.

Inflation tells us that the clusters and groups of galaxies today originated as small random quantum fluctuations in the density of the universe. Some regions were above average in terms of density, and some were below average. As gravity works on these regions, the denser than average regions become denser still by drawing material toward them. Meanwhile, the below-average regions expand at a faster than average rate and thin out. Since these fluctuations are random, the high-density region of space must have a geometry similar to that for the low-density region.

A sponge has this type of geometry as well. A sponge is multiply connected, with many holes. Its insides and outsides are similar. If high- and low-density regions of the universe were originally produced by random quantum

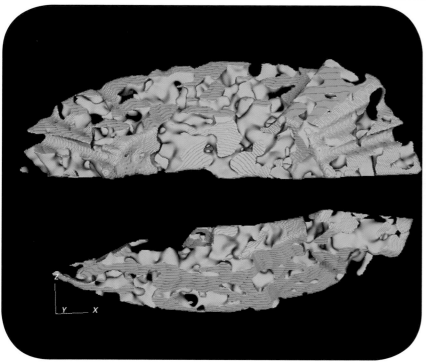 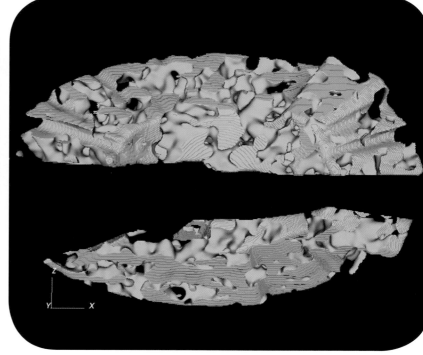

fluctuations as predicted by inflation, they should be spongelike. As gravity works on these regions, the denser regions should get denser and the low-density regions should get emptier, but their geometry should remain spongelike. Gott, Adrian Melott, and Mark Dickinson predicted that this should lead to a spongelike pattern of galaxy clustering today, in which dense clusters of galaxies are connected by filaments (or chains) of galaxies and empty voids are connected to each other by low-density tunnels containing few galaxies. Large computer simulations of gravitational clustering by Changbom Park and Juhan Kim have supported this prediction. If inflation is right that random quantum fluctuations were the seeds that started galaxy clustering, then we should see a spongelike pattern of galaxy clustering today.

We can test this by examining large 3-D galaxy surveys. In the (cross-eyed) stereo view, we show recent results Gott and Clay Hambrick and their colleagues

have obtained from a Sloan Digital Sky Survey 3-D data set that included more than 400,000 galaxies. Two thick slices are shown. These surveyed regions are fan-shaped thick slices with Earth located in front of them. You are looking from a position farther in front. The high-density regions (where the counts of galaxies are high) are shown as solid green, while the low-density regions are shown as empty. You can clearly see the spongelike distribution, which is exactly as predicted. The bottom slice is the equatorial slice shown in the Map of the Universe. The Sloan Great Wall is a long filament at the bottom of this slice that stretches from near the center to the far right of the slice.

Three-dimensional view of the spongelike pattern of galaxy clustering known as the cosmic web. This supports the theory of inflation, in which such structures began as random quantum fluctuations in the very early universe.

← G76

5,370 ly

Andromeda Galaxy (M31)

← G64

M110

GALAXIES COMPARED

M87

Sombrero

Whirlpool

Milky Way

Andromeda

M81

M82

59,300 ly

Step Nine

AT THIS SCALE, we can see the whole Local Group and its most prominent members, the Milky Way, M31, and M33. All are spiral galaxies. We are a sparse, small group. The Local Group started as a region of slightly higher than average density in the universe. Because of this excess matter, gravity continued to slow its expansion relative to the rest of the universe, until it stopped expanding, and M31 and the Milky Way, which had already formed, started to fall back toward each other. They should collide in another 3.1 billion years.

FYI We suspect that dark matter is made up of weakly interacting massive particles, or WIMPS, which are more massive than protons.

The Coma cluster is a rich cluster. In the center of the Coma cluster are two enormous elliptical galaxies. James Gunn and I (Gott) estimated the mass of each to be more than 10^{13} solar masses. We noted that, as intergalactic gas fell into the cluster from different directions, the gas would become heated to a temperature of 70 million kelvin. This hot gas, observable in x-rays, strips spiral galaxies passing through it of their cool gas, robbing them of their star-forming potential. In 1933 Fritz Zwicky measured the velocities of the galaxies in the Coma cluster and discovered that the mass required to hold the cluster together was much greater than the total mass of the galaxies; the cluster must be held together by additional nonluminous mass—dark matter.

The Perseus cluster is another great cluster of galaxies held together by dark matter. In the Bullet cluster two clusters have collided and passed through each other. The total mass, dominated by dark matter, causes small gravitational lensing distortions of the shapes of faint background galaxies as it bends light beams coming from them to us. This allows us to map the dark matter (shown in blue).

As mapped by the Chandra X-ray satellite, the hot gas can be seen as two red lumps that have collided and passed through but have been slowed by the collision. The dark matter and galaxies have passed through unimpeded to form two lumps to the left and right of the hot gas. The dark matter (providing most of the gravity and gravitational lensing) is not distributed like the ordinary matter (gas plus luminous galaxies). Weakly interacting particles would behave like this.

To the left is a portion of the cosmic microwave background sphere. The cold spots in the cosmic microwave background (shown in blue) are where the density is greatest and photons have lost energy climbing out toward us. These denser than average regions will eventually collapse to form clusters of galaxies. Hot spots shown in red are regions of lower than average density that will expand faster than average to produce giant voids.

When we look out in space, we look back in time, as illustrated in the 3-D picture opposite. The big bang explosion at the bottom is powered by a short period of hyperactive expansion known as inflation. Everything expands outward from the big bang with time. The black, teardrop-shaped surface shows what we can see when we look out now. The horizontal directions represent space, whereas the vertical direction represents time. We are at the top of the image (where the "Here and Now" arrow points). Green spots are galaxies, and orange spots are quasars.

We are showing them at the place and time they emitted the light we see today. But they were thrown outward

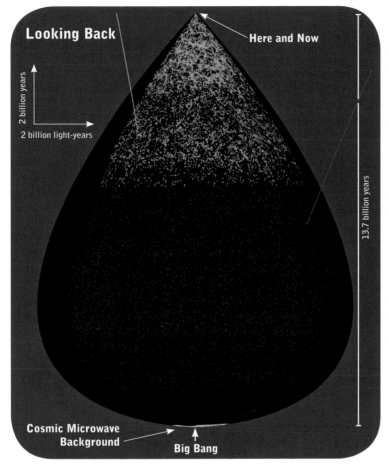

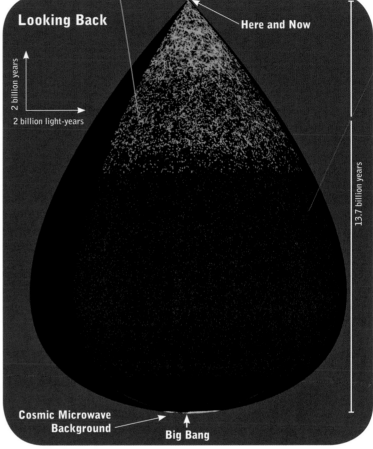

by the big bang explosion, so they are farther "out" horizontally today (at what we call their co-moving distance) than they were at the time they emitted the light we see today. The path (world line) of a typical galaxy, as it continues to move outward from where we see it to where it actually is today, is shown as a solid green line. A similar world line (orange) is shown for one typical quasar.

Looking out farther and farther through curved space-time, we are looking back to epochs when the universe was smaller. If we could look back far enough in any direction, we would see a "south pole" of microscopic size and extreme density—the big bang itself (at the very bottom). But our view is limited by the fact that the universe is no

longer transparent at early epochs. It is like looking into a fog bank. The earliest we can see is an epoch 380,000 years after the big bang, when the universe was 1,090 times smaller than it is today. The physical size of the cosmic microwave background sphere (83.8 million light-years in diameter, shown on the following pages) is small compared with its distance from us, because it is near the south pole of space-time—the big bang—when the universe was small.

A 3-D space-time diagram showing what we see when we look out. When we look out in space we look back in time to an epoch when the universe was small. Time is vertical in this diagram. The other two dimensions in the picture represent the dimensions of space.

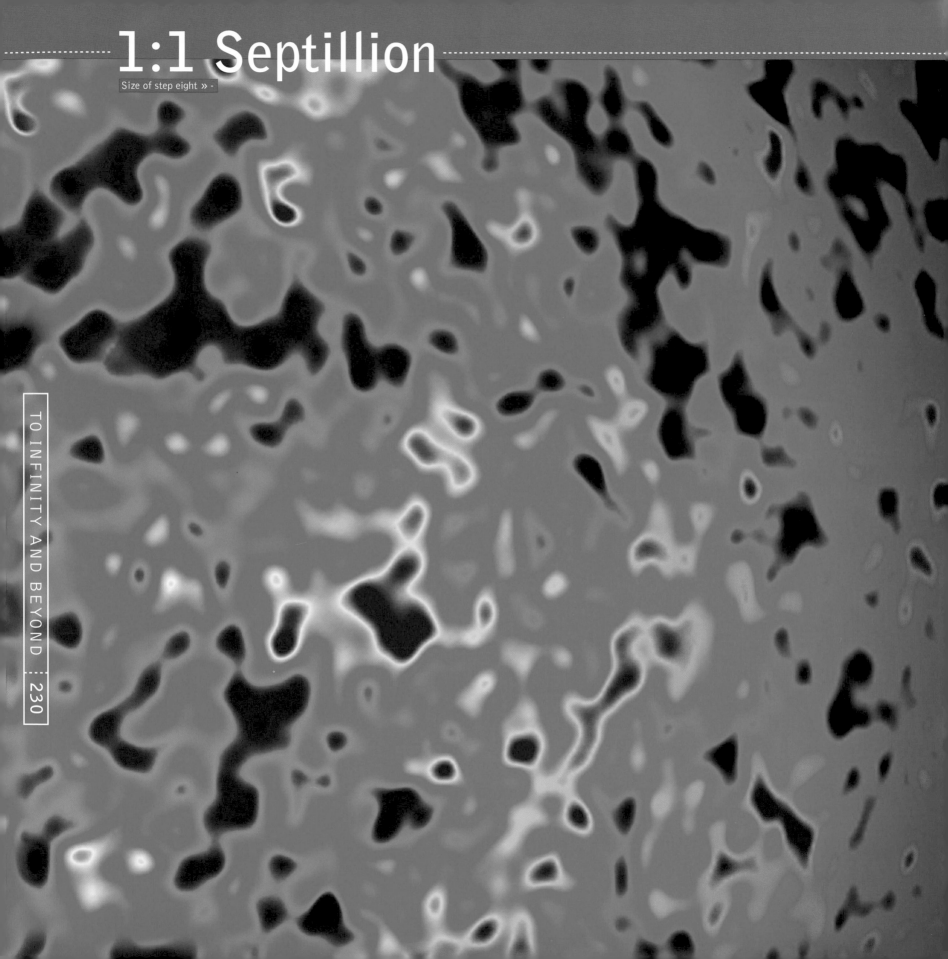

Cosmic Microwave
Background

M31
M33

Milky Way

Perseus Cluster

Local Group

Coma Cluster

Bullet Cluster

Step Ten

ONE LAST JUMP by a factor of a thousand, and we can see the extent of the entire visible universe—everything we can see. We are looking here at the equatorial slice of the Sloan Survey containing 126,594 galaxies and quasars. It is a cross-sectional slice of the universe extending outward from Earth's Equator. Earth is in the center of the picture. Galaxies are shown as green dots, and quasars as orange dots. The two large, blank regions are zones of avoidance, where our galaxy blocks the distant view. The scale shows the look-back-time distance in billions of light-years.

When we look out in space, because of the finite velocity of light, we look back in time. A galaxy five billion light-years away we see as it was five billion years ago. We can see out to a radius of just 13.7 billion light-years in any direction, because the universe began in a big bang explosion 13.7 billion years ago. The farthest thing we can see is the cosmic microwave background radiation left over after the big bang, which encircles the visible universe.

Earth is at the center of the visible universe. This does not mean that we are in a special location. If you look out from the top of the Empire State Building in New York City, the region you can see, out to the horizon, is circular and centered on the Empire State Building. Looking out from the top of a different building, you would see a different circular region—one centered on it. An observer in a distant galaxy would see a different visible region of the universe, one centered on his galaxy instead of ours. Most of the galaxies visible are less than 5 billion light-years away, while most of the quasars are between 5 and 12 billion light-years away. Near Earth, many voids and walls of galaxies are visible. These also appear in the Map of the Universe shown on pages 123-126.

The Sloan Great Wall (highlighted by a detailed, same-size inset that points to its location in the map) is the largest structure we have found in the universe so far. Its length stretches 1.37 billion light-years, one-tenth the radius of the visible universe.

FYI We are now observing a scale a billion, billion, billion times larger than Buzz Aldrin's footprint. And that footprint is a billion times larger than a hydrogen atom.

The visible universe is very large because gravity is such a weak force. In 1961, physicist Robert Dicke pointed out that it is no accident we live about one stellar main-sequence lifetime after the big bang—after some stars have died (to make the carbon needed for life) but before all the stars have burned out, making it too cold for life. If gravity were stronger, main-sequence lifetimes of stars would be shorter, we would live closer in time to the big bang, and the radius of the visible universe would be smaller. Since gravity is weak, we carbon-based life-forms, orbiting our main-sequence star, are treated to a truly grand view.

Just think: After the end of inflation, the process that produced the big bang explosion 13.7 billion years ago, all the matter we can see in the visible universe today was still inside a region smaller than Buzz Aldrin's footprint.

A cross-sectional map of the visible universe. Sloan Survey galaxies (green) and quasars (orange) are plotted at their look-back-time distances—black wedges are regions obscured by our galaxy and not covered by the Sloan Survey. Around the perimeter is the cosmic microwave background.

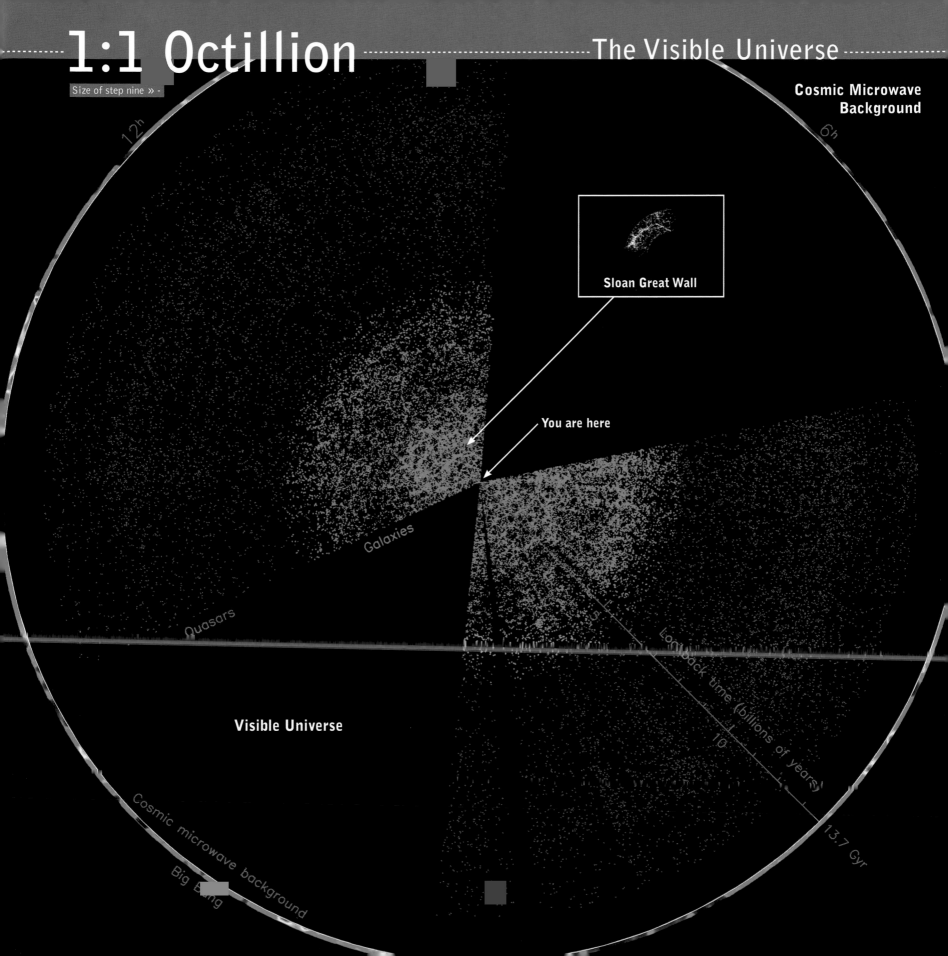

Cosmic Microwave Background

Sloan Great Wall

You are here

Galaxies

Quasars

Visible Universe

Lookback time (billions of years)

10

13.7 Gyr

Cosmic microwave background

Big Bang

12ʰ

6ʰ

And Beyond

WHAT'S BEYOND the visible universe? Galaxies and more galaxies for a long way beyond the boundary of what we can see. Measurements of the cosmic microwave background by the WMAP satellite have enabled us to survey accurately the visible universe. The results tell us that a slice through the universe at the present epoch appears flat, like a sheet of paper. Keep extending a flat sheet of paper, and it forms an infinite flat plane. That might suggest the universe is infinite, with an infinite number of galaxies. Actually, it just means that the universe is much larger than the part we can see. Manhattan looks flat but actually is part of the finite, but large, curved surface of Earth. The universe could be almost any shape, but if it is large enough, any piece of it will look flat. All we really know for sure is that the true size of the universe is much larger than 13.7 billion light-years. Alan Guth's theory of inflation predicts this, but how much bigger is it?

Physicist Andrei Linde's theory of chaotic inflation suggests that our universe today may be $10^{40,000}$ times larger than the part we can see. Our universe forms in a high-density inflationary sea of dark energy. This dark energy has a very large negative pressure, and the repulsive gravitational effects of this pressure start a runaway expansion. This sea doubles in size every 10^{-37} seconds. Such a sequence (1, 2, 4, 8, 16, 32, 64, 128, 256, 512, 1024, . . .) blows up quickly. Inflation in our region ends when the dark energy converts into thermal radiation (the hot big bang) and the explosive expansion starts to slow down.

In my (Gott's) theory of bubble universes we are just one bubble in this expanding sea. Our bubble expands at nearly the speed of light and grows to infinite size, eventually creating an infinite number of galaxies inside. Beyond our own bubble universe is the inflating sea, which forms an infinite number of other bubble universes—a *multiverse*.

But there's more. Linde has shown that inflationary regions can give birth to other inflationary regions through quantum fluctuations, like branches growing off a tree. Each branch grows up to be as big as the trunk and sprouts branches of its own. The process, once started, continues forever, producing an infinite fractal tree with an infinite number of branches.

FYI According to Linde's theory, it would take over **10,000** more pictures, each covering a scale a thousand times larger than the one before, to capture our universe in its full glory.

Our universe is just one of many on one of these branches. We will never see these other branches, because the space separating us from them is expanding so fast that light can never cross it. Where did the trunk come from? Perhaps it simply popped into existence by a process called quantum tunneling, as proposed by Alex Vilenkin. Or perhaps it formed from a time loop, as proposed by me (Gott) and Li-Xin Li, when a branch looped back in time to grow up to be the trunk. We simply don't know. Superstring theorists Paul Steinhardt and Neil Turok have suggested that our universe is a three-dimensional membrane floating in a ten-dimensional space and that the big bang occurred when our membrane collided with another.

If our universe is a bubble, it is infinite in extent. Beyond our universe, myriad other universes in the multiverse branch off in the distance, farther than the eye can ever see.

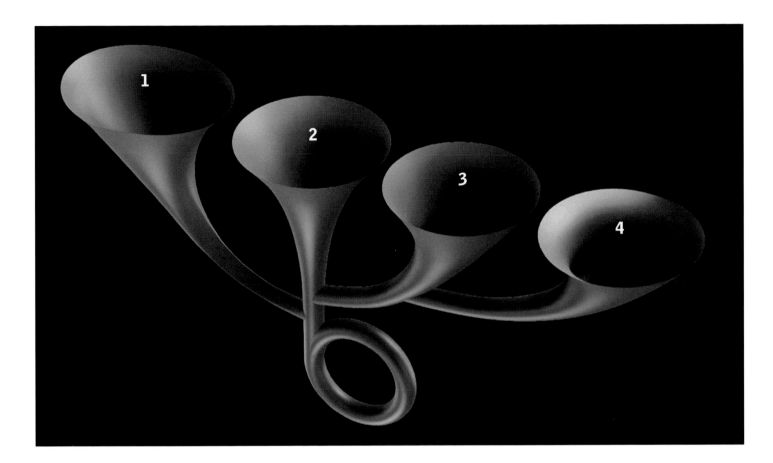

The space-time diagram above shows the Gott-Li model for the creation of multiple universes from a time loop. Each funnel represents an inflating universe that is growing larger with time—we show one dimension of space (the circumference of the funnel) and the dimension of time (running upward out the funnel). We label four universes 1, 2, 3, and 4, from left to right. Universe 2 gives birth to universes 1 and 3—they are its children. Universe 3 gives birth to universe 4. So universe 4 is the grandchild of universe 2. Each branch grows to infinite size and sprouts branches of its own. We live in one of the many branches. Physicist Andrei Linde has shown that, once started, this branching process will continue forever, creating an infinite fractal tree of universes.

But where did the original trunk come from? Alex Vilenkin, Stephen Hawking, and James Hartle suggest it simply pops into existence. But Gott and Li have proposed that universe 2 was its own mother, giving birth to a branch that circled back in time and grew up to be the original trunk.

The Gott-Li model represents just one of a number of speculative possibilities being explored by physicists for the origin of our inflating universe. Studies of the cosmic microwave background suggest that inflation is likely, but details of how it got started have yet to be worked out.

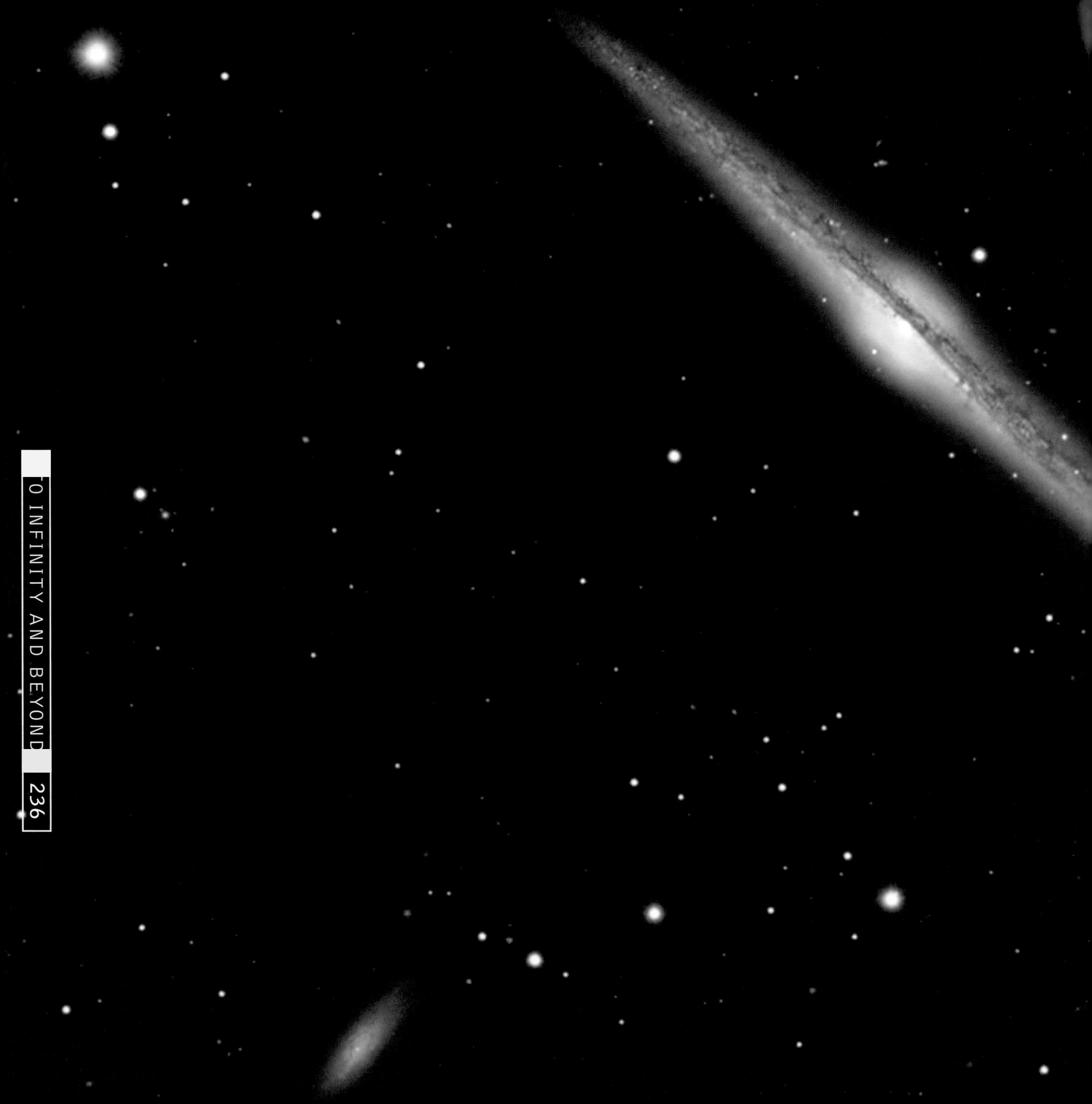

First discovered in 1785 by Sir William Herschel, who also discovered Uranus, NGC 4565 is a popular object for amateur astrophotographers such as Dietmar Hager, who took this beautiful picture.

AFTERWORD

I BECAME A SERIOUS amateur astronomer about 12 years ago. A friend, Kirk Alexander, had invited me to a weekend star party at a dark-sky location not far from where I live in New Jersey. I went and I was blown away by the things that I saw that evening. I've been hooked on astronomy ever since.

I bought my first telescope in 1999. It is a reflector telescope with a mirror that is only 3.5 inches in diameter. I quickly discovered one disappointing fact: Light pollution in New Jersey is brighter than most of the faint nebulae and galaxies that I was hoping to see. The moon and most planets are beautiful and bright; they can be viewed even from the brightest urban areas. In fact, they can even be observed during the day. Stars are also viewable from urban areas, and open star clusters, such as the Beehive cluster (M44), look great as do colorful binary star systems, such as Albireo. But only a handful of the brightest nebulae and galaxies could be seen in my little telescope.

One might think that all I needed was a larger telescope. Surprisingly, that wasn't the answer. A larger telescope can make faint astronomical objects brighter, but it also makes the glow from the light pollution brighter to the same degree. Fortunately, some products of modern technology have brought these "faint fuzzies" back into the realm of backyard (or, in my case, driveway) astronomy.

First of all, there are so-called light-pollution filters. These filters exploit the fact that much of the light pollution bathing us comes from sodium vapor lamps and mercury vapor lamps. These lamps emit light consisting of only a few specific wavelengths. A light-pollution filter blocks these wavelengths but lets everything else through. The result is that most of the light from the object we want to see gets to our eyes while much of the light pollution gets blocked.

A related solution exploits the fact that most of the beautiful nebulae we would like to observe are so-called emission nebulae, which means that they emit light in only a few wavelengths. Most of the light is emitted with a deep red wavelength of 6563 angstroms—this light is called hydrogen-alpha (or just H-alpha). The second most predominant wavelength is 5007 angstroms—a blue-green color called oxygen-III (or just O-III). Although these narrow-band filters can be used visually, they are mainly used for astrophotography. The picture of the Helix Nebula, shown on the next page, was taken with my 4-inch refractor using an H-alpha and an O-III filter.

In addition to narrow-band filters, modern telescopes, even the most modest ones, are mounted on highly accurate computer-controllable mounts. These mounts with their precise tracking capabilities make it possible for backyard astronomers to take long-exposure photographs with beautiful pinpoint stars. Furthermore, modern digital cameras are amazingly sensitive and when coupled to a laptop computer provide quick feedback as to how the imaging session is going. These technological advances, some of them only a decade or so old, allow amateur astronomers with modest equipment to take astrophotographs that rival, and in some cases surpass, those taken by the world's largest telescopes only a generation ago.

Why Do It?

One of the arguments against spending the time to take pictures yourself is that there are already excellent

The moon at first quarter, taken by Vanderbei from his backyard in New Jersey with a point-and-shoot camera pointed into the eyepiece of his 3.5-inch reflector telescope.

pictures in books and on the Internet. So why invest the time to do it? There are several good answers. First, it is fun and rewarding to take pictures yourself. Consider any famous tourist site such as the Grand Canyon or Niagara Falls. Millions of pictures have already been taken of these places, yet every visitor wants to take a picture of his or her own. Second, one learns something from going to the effort of taking the picture. As the old saying goes, "No pain, no gain." For example, you might ask yourself: Which object is brighter in the sky, the Crab Nebula (M1) or the Dumbbell Nebula (M27)? Most people don't know. You could look it up. But you'd likely soon forget the answer. If you have imaged both objects, however, there is no chance that you will ever forget which one is brighter—M27. The same applies to angular scales. Which appears bigger in the sky, the Eagle Nebula (M16) or the Rosette Nebula (NGC 2237)? The angular scales chapter in this book gives you an appreciation for angular scales. If you have spent

hours setting up and taking a picture of both, then you know which appears larger—the Rosette Nebula. Such knowledge becomes part of your experience of observing.

Some astronomical events are transitory, and direct observing gives you a front row seat to the action. Comets appear, hang around for a few months, and then are gone. Some put on spectacular displays. In 1994, Comet Shoemaker-Levy 9 plunged into Jupiter and left a scar that was visible for many weeks. Recently a comet passed by globular cluster M3. This provided an interesting and unusual juxtaposition of two very different objects (shown opposite).

In the mid-1700s, Charles Messier (1730–1817), who was looking for comets, made a list of 103 objects in the night sky. Because the objects did not move from one night to the next, they were evidently not comets. These objects annoyed him, but much later astronomers realized that these objects were quite interesting themselves. The third item on his list, M3, is a globular cluster, a vastly larger and more distant object than a comet. So it was remarkable when a recent comet (C/2006 VZ 13) passed close enough to M3 to enable a picture of both to be taken simultaneously.

Sizing Up the Universe for Yourself

Amateur astronomers not only can take beautiful pictures; they also can fairly easily reproduce some of the experiments described in Chapter 3 to measure the distances to, and sizes of, astronomical objects. A few years ago fellow amateur astronomer Rus Belikov and I measured the distance from Earth to an asteroid, Prudentia, by taking pictures of it just after sunset, then just before sunrise, and again just after sunset on the second night. Over the course of that 24-hour period, the asteroid's position in the sky (relative to the background stars) moved. Using

the two evening observations, an accurate time log, and assuming a constant motion against the stars, we were able to predict where the asteroid should be in the sky when we took the early morning picture. As expected, it was displaced from this predicted location. The reason is that we were imaging from the surface of Earth, and Earth rotates. Our viewing point on Earth had shifted and hence we were displaced by about one Earth diameter when we took the morning picture. The displacement turned out to be about ten arc-seconds, which was easy to measure even with a small telescope. From our pictures, we got a distance that was within one percent of the correct distance to Prudentia. A planetarium program told us the distance to Prudentia in astronomical units (AU) at the time of our observations and therefore we were able to calculate how many Earth diameters make 1 AU.

Hertzsprung-Russell diagrams, discussed on page 208, played an important role in measuring the distance to open clusters and globular clusters. It takes only modest computer programming skills to extract from amateur images the information needed to construct such diagrams.

I hope you have enjoyed reading this book as much as I have enjoyed writing it, and I hope we have inspired some of you to participate actively in amateur astronomy. Today, amateur astronomers are assisting professional astronomers in discovering and characterizing extra-solar planets. I'm sure many of you already have the skills needed to participate. For anybody with an interest, amateur astronomy provides lasting pleasures and an opportunity to make a real contribution to the advancement of the science.

—*Robert J. Vanderbei*

Vanderbei's images of the Helix Nebula (opposite) and comet Linear C/2006 VZ 13 and globular cluster M3 together in the sky (above).

Illustrations Credits

All illustrations by Theophilus Britt Griswold unless otherwise noted.

Key: RJV = Robert J. Vanderbei
           JRG = J. Richard Gott

1, NASA, JPL-Caltech, and R. Hurt (SSC); 2-3, NASA, JPL-Caltech, and R. Hurt (SSC); 4-5, NASA and JPL-Caltech; 6-7, NASA, ESA, N. Smith (University of California, Berkeley), and The Hubble Heritage Team (STScI/AURA); 8-9, RJV; 10-11, RJV; 12, RJV; 13, JRG; 15, RJV; 17, RJV, and NASA (HST); 20-21, Laurent Laveder; 23, Alaska Stock Images/NationalGeographicStock.com; 25, Jim Richardson; 26-27, RJV based on data from Main Sequence Software; 29, Stefan Seip/NASA; 30, Richard Cummins/Corbis; 31 (LE), RJV; 31 (RT), NASA; 33, RJV; 35, RJV, based on data from Palomar Online Sky Survey; 36-37, Moon, RJV, Andromeda, Robert Gendler; 38-39, Mars: NASA; Saturn: NASA; Hubble Ultra Deep Field; NASA (STScI); Moon: NASA; 41, Betelgeuse: NASA; Other Stars: RJV; Apollo 11 on Moon: NASA Lunar Reconnaisance Orbiter; 43, Stars: RJV; Footprint: NASA; 44-45, Stephen L. Alvarez; 46 (UP), Wes Colley; 46 (LO), Wes Colley; 47, RJV based on data from Main Sequence Software; 54, M. Aschwanden, LMSAL, TRACE, and NASA; 55, Sun: SOHO; Moon: RJV; Mercury: NASA; Venus: NASA; Mars: NASA; Jupiter: RJV; Saturn: RJV; Uranus: NASA; Neptune: NASA; Pluto: NASA; Eris: NASA; HST: NASA; Ceres: NASA; Brightest Quasar: NASA; Stars: RJV; 57, RJV; 58, Blinking Planetary: B. Balick, J. Alexander, et al., NASA; Bode's Galaxy: RJV; Bubble: RJV; Butterfly Nebula: RJV; Cat's Eye: RJV; Cave: Robert Gendler; Cocoon: Julie and Jessica Garcia, Adam Block, NOAO/AURA/NSF; Double Cluster: Michael Fulbright (MSFAstro.net); Elephant Trunk: RJV; Helix: Robert Lupton and the SDSS Consortium; Integral Sign: Bonnie Fisher and Mike Shade, Adam Block, NOAO/AURA/NSF; Little Dumbbell: RJV; North American Nebula: RJV; Owl: RJV; Pacman: RJV; Pelican: RJV; Whirlpool: RJV; 60-61, Robert Lupton, Sloan Digital Sky Survey; 62, Beehive: RJV; California: CalTech, Palomar Observatory, Digitized Sky Survey; Cone: RJV; Crab RJV; Eskimo: RJV; Hind's Variable: 2MASS/IPAC; Horsehead: RJV; Hubble's Variable Nebula: NASA/ESA and The Hubble Heritage Team (AURA/STScI); Intergalactic Wanderer: Doug Matthews and Charles Betts, Adam Block, NOAO/AURA/NSF; Medusa: Al and Andy Ferayorni, Adam Block, NOAO/AURA/NSF; Orion Nebula: RJV; Pleiades: NASA; Rosette: RJV; Witch Head: NASA; 64-65, NASA; 66, Antennae: NASA, ESA, and the Hubble Heritage Team STScI, AURA-ESA, Hubble Collaboration; Blackeye: NASA/ESA and The Hubble Heritage Team (AURA/STScI); Cocoon: Michael Gariepy, Adam Block, NOAO/AURA/NSF; Coma Cluster: Robert Gendler; Copeland's Septet: Neil Jacobstein, Adam Block, NOAO/AURA/NSF; Ghost of Jupiter: Bruce Balick (U. Washington) et al., HST, NASA; Leo Trio: RJV; M87: NASA and The Hubble Heritage Team (STScI/AURA); Markarian's Chain: R. Gilbert, J. Harvey et al. (SSRO); Mice: NASA, Holland Ford (JHU), the ACS Science Team and ESA; NGC 4565: RJV; QSO 3C 273: NASA; Siamese Twins: Bill and Marian Wallace, Adam Block, NOAO/AURA/NSF; Sombrero: RJV; Spindle: NASA, ESA, and The Hubble Heritage Team STScI/AURA; Sunflower: RJV; 68-69, NASA, STScI, AURA, and The Hubble Heritage Team; 70, Albireo: RJV; Sgr A: NASA, CXC, MIT, F. K. Baganoff et al.; Coat Hanger T. Credner & S. Kohle, AlltheSky.com; Crescent Nebula: RJV; Dumbbell Nebula: RJV; Eagle Nebula: RJV; Hercules Cluster: RJV; Lagoon Nebula: RJV; M5 RJV; Rho Ophiuchi Nebula: J. Ballauer and Phil Jones, NOAO; Ring Nebula: RJV; Seyfert's Sextet: NASA; Swan Nebula: RJV; Snake Nebula: Tom McQuillan, Adam Block, NOAO/AURA/NSF; Trifid Nebula: RJV; Veil Nebula: RJV; Wild Duck Cluster: RJV; 72-73, NASA, JPL, and S. Stolovy (SSC/Caltech); 74, Andromeda: Robert Gendler; Blue Snowball: Balick, Hajian, Terzian, Perinotto, Patriarchi (STScI); Cartwheel: NASA/STScI; Einstein's Cross: J. Rhoads, S. Malhotra, I. Dell'Antonio (NOAO), WIYN/NOAO/NSF; Helix: NASA, ESA, and C. R. O'Dell (Vanderbilt University); M2: RJV; M15: RJV; M74: Gemini Observatory, GMOS Team; M33: RJV; Sculptor Galaxy: Atlas Image courtesy of 2MASS, UMass, IPAC-Caltech, NASA, NSF; Stephan's Quintet: N. A. Sharp, NOAO/AURA/NSF; 76-77, NASA, ESA, and The Hubble Heritage Team; 78, Ant: NASA/JPL; Circinus Galaxy: NASA/A.Wilson et al.; Coal Sack: Don Pettit, ISS Expedition 6, NASA; Eta Carinae: Jon Morse (University of Colorado) and NASA/ESA; Gum Nebula: Christopher J. Picking; Jewel Box: Jack Harvey, SSRO/SKYNET; Large Magellanic Cloud: Chris Smith, CTIO/NOAO; Omega Centauri: Bernd Flach-Wilken and Volker Wendel; Small Magellanic Cloud: A. Nota (ESA/STScI) et al., ESA, NASA; Tarantula: Spitzer Space Telescope; Vela SNR: Digitized Sky Survey/ESA/ESO/NASA FITS Liberator; 80-81, NASA, JPL-Caltech, and T. Pyle (SSC); 85, RJV; 93, RJV; 97, Azcolvin (CC-BY-SA 3.0); 98-99, RJV; 100, NASA; 101, Robert Gendler; 103, RJV; 104-105, 2MASS; 107, Subaru Telescope, NAOJ; 108-109, JRG; 110-111, NASA, ESA, and The Hubble Heritage Team (STScI/AURA); 113, JRG and Mario Jurić; 115, Lorne Hofstetter and JRG; 116-117, Lorne Hofstetter and JRG; 118, The New Yorker Magazine, Inc./Original Artwork: Saul Steinberg Foundation; 119 (LE), RJV; 119 (RT), RJV; 130, NASA; 131, NASA/W. Liller; 133, RJV; 135, Wes Colley and NASA; 136-137, Photovideostock/iStockphoto.com; 139, NASA/Mariner 10; 140-141, Mercury: NASA; Venus: NASA; Earth: NASA; Moon: RJV; Mars: NASA; 142-143, RJV; 144, NASA; 145, NASA; 146 (LE), Moon front: RJV; Moon back: RJV; Earth map: RJV; 147, JRG, Mugnolo, Colley; 148-149, NASA, U.S. Geological Survey; 149 (UP), U.S. Geological Survey; 150-151, NASA; 152 (LE), RJV; 152-153, Wes Colley; 154 (LO), Wikipedia; 154-155, NASA/JPL; 156-157, NASA/JPL; 158, Alaska Stock Images/NationalGeographicStock.com; 159, NASA; 161, The Exploratorium/NASA; 162-163, NASA, SOHO; 164, RJV; 165, NASA; 166-167, NSO, Friedrich Woeger, Chris Berst, Mark Komsa; 166 (LO), RJV; 168 (LO), RJV; 168-169, David Goldberg; 170 (LO), NOAA; 170-171, NASA; 172, NASA; 173, NASA; 174-175, Saturn Aurora: Cassini VIMS Team, JPL, ESA, NASA; Aurora Australis: NASA's IMAGE satellite and NASA's Blue Marble; 177, NASA, Cassini; 178-179, Moon: RJV; All others: NASA; 180-181, Hawaii: NASA/Corbis; Lake Michigan: NASA; Olympus Mons: NASA; Io Volcano: Galileo Project, JPL, NASA; Methane lakes on Titan: NASA, Cassini; 182-183, NASA; 184-185, NASA, ESA, and The Hubble Heritage Team STScI/AURA; 187, NASA, ESA, and The Hubble Heritage Team STScI/AURA; 191, NASA; 192-193, NASA; 195, NASA; 196-197, NASA; 199, Andrew Hamilton; 200-201, NASA; 203, NASA Cassini; 204-205, Saturn: NASA/JPL; Neptune: NASA/JPL; Uranus: Keck; Jupiter: NASA/JPL; Io: NASA/JPL; Europa: NASA/JPL; Ganymede: NASA/JPL; Callisto: NASA/JPL; Sun: SOHO; Titan: NASA/JPL; Mars: NASA/JPL; Earth: NASA/GSFC; Moon: RJV; Venus: RJV; Mercury: USGS; Proxima Centauri: RJV; 206-207, RJV; 209, RJV; 210-211, RJV; 213, RJV; 214-215, Vega dust disk: NASA/JP; Solar System orbits: RJV; 217, RJV; 218-219, M13: RJV; Horsehead: RJV; Ring: RJV; Dumbbell: RJV; Orion: RJV; Crab: RJV; Eskimo: RJV; Eagle: NASA/SScI; SN 1987a: NASA/JPL; 220-221, RJV; 223, Clay Hambrick and JRG; 224-225, Robert Gendler; 226-227, Robert Gendler; 229, RJV; 230-231, CMB: Wes Colley and NASA; Perseus Cluster: SDSS; Coma Cluster: SDSS; Bullet Cluster: NASA/Chandra; 233, CMB: JRG, Mario Jurić, Wes Colley and NASA, and RJV; Sloan Great Wall: SDSS; 235, Christopher Sloan and David Fierstein; 236-237, Dietmar Hager; 239, RJV; 240, RJV; 241, RJV.

GATEFOLD

Front, Universe Map: RJV, JRG, Mario Jurić, and Theophilus Britt Griswold; Hubble: NASA/STSci; WMAP: NASA/WMAP Science Team; Mars: NASA/STSci; Mercury: Johns Hopkins University/APL; Sun: NASA ESA SOHO EIT; Venus: NASA/USGS/JPL; Jupiter: NASA/JPL/Space Science Institute; Saturn: NASA/STSci; Uranus: Keck Observatory/Lawrence Sromovsky/UWM; Neptune: NASA/Voyager 2; Dumbbell: RJV; Orion Nebula: RJV; Crab Nebula: RJV; M33: Thomas V. Davis; M31: Robert Gendler; Whirlpool: RJV; M81: RJV; Sombrero: RJV; Coma Cluster: NASA/STSci.

Back (L to R), CMB: NASA, Bennett, et al. 2003; Mice: NASA/ESA; Eskimo: Andrew Fruchter (STScI) et al., WFPC2, HST, NASA; Earth: Wes Colley and NASA.

Acknowledgments

THE AUTHORS wish to thank Neil deGrasse Tyson, Rus Belikov, and Chuck Allen for reading early drafts of the manuscript and making helpful suggestions. We thank our agent, Jeff Kleinman, of Folio Literary Management. We also thank Andrew Hamilton, Rob Gendler, Dietmar Hager, Clay Hambrick, Changbom Park, David Weinberg, Michael Vogeley, Juhan Kim, Yun-Young Choi, and Robert Lupton for their memorable images.

We owe a special debt to Mario Jurić, Wes Colley, Lorne Hofstetter, and Dave Goldberg for their creative collaboration in producing the maps and images that are central to the concept for this book. And, for translating our vision into a wonderful National Geographic book accessible to a broad readership, editor Garrett Brown and art director Cinda Rose have our lasting gratitude.

We dedicate this book to the women in our lives— Lucy Pollard-Gott, Elizabeth Gott, Krisadee Vanderbei, Marisa Vanderbei, Diana Vanderbei, Marilyn Schuitema, and Marjorie Crosby Gott (1912–2004; former president of the Garden Club of Kentucky and an appreciator of beauty)—and to anyone who takes pleasure in the beauty of the universe.

About the Authors

J. RICHARD GOTT III is professor of astrophysics at Princeton University, where he received his doctorate. He is the author of *Time Travel in Einstein's Universe* and has written for *Time, Scientific American, New Scientist,* and other publications. He and his work have been profiled in *The New Yorker, Time, Newsweek, National Geographic,* and the *New York Times.* His measurement of the Sloan Great Wall of galaxies was entered into *Guinness World Records 2006* as the "largest structure in the universe."

For more information about Gott's work in astrophysics or time travel, or about Vanderbei's astrophotography or practical observation tips, please visit http://www.sizinguptheuniverse.com.

ROBERT J. VANDERBEI is professor and chair of the Department of Operations Research and Financial Engineering at Princeton University. He received his doctorate in applied mathematics from Cornell University and contributed to many of the leading design concepts for NASA's Terrestrial Planet Finder space telescope. He is also known for his "Purple States of America Map," reprinted in *Time* and *Newsweek,* which showed how votes in the presidential elections of 2000 and 2004 were divided county by county. As an amateur astronomer, Vanderbei has taken, from his own backyard, high-quality images of astronomical objects that rival the best images from our greatest observatories.

Index

Boldface indicates illustrations

SIZING UP THE
UNIVERSE
THE COSMOS IN PERSPECTIVE

J. Richard Gott III and Robert J. Vanderbei

Published by the National Geographic Society

John M. Fahey, Jr., *President and Chief Executive Officer*
Gilbert M. Grosvenor, *Chairman of the Board*
Tim T. Kelly, *President, Global Media Group*
John Q. Griffin, *Executive Vice President; President, Publishing*
Nina D. Hoffman, *Executive Vice President;*
    President, Book Publishing Group

Prepared by the Book Division

Barbara Brownell Grogan, *Vice President and Editor in Chief*
Marianne R. Koszorus, *Director of Design*
Lisa Thomas, *Senior Editor*
R. Gary Colbert, *Production Director*
Jennifer A. Thornton, *Managing Editor*
Meredith C. Wilcox, *Administrative Director, Illustrations*

Staff for This Book

Garrett Brown, *Editor*
Cinda Rose, *Art Director and Designer*
Cameron Zotter, *Designer*
Adrian Coakley, *Illustrations Editor*
Judith Klein, *Production Editor*
Lisa A. Walker, *Production Manager*
Robert Waymouth, *Illustrations Specialist*
J. Richard Gott and Robert J. Vanderbei, *Cartographers*
Theophilus Britt Griswold, *Illustrator*

Manufacturing and Quality Management

Christopher A. Liedel, *Chief Financial Officer*
Phillip L. Schlosser, *Vice President*
Chris Brown, *Technical Director*
Nicole Elliott, *Manager*
Rachel Faulise, *Manager*
Robert L. Barr, *Manager*

The National Geographic Society is one of the world's largest nonprofit scientific and educational organizations. Founded in 1888 to "increase and diffuse geographic knowledge," the Society works to inspire people to care about the planet. It reaches more than 325 million people worldwide each month through its official journal, *National Geographic,* and other magazines; National Geographic Channel; television documentaries; music; radio; films; books; DVDs; maps; exhibitions; school publishing programs; interactive media; and merchandise. National Geographic has funded more than 9,000 scientific research, conservation and exploration projects and supports an education program combating geographic illiteracy. For more information, visit nationalgeographic.com.

For more information, please call 1-800-NGS LINE
(647-5463) or write to the following address:

National Geographic Society
1145 17th Street N.W.
Washington, D.C. 20036-4688 U.S.A.

Visit us online at www.nationalgeographic.com

For information about special discounts for bulk purchases, please contact National Geographic Books Special Sales: ngspecsales@ngs.org

For rights or permissions inquiries, please contact National Geographic Books Subsidiary Rights: ngbookrights@ngs.org

Library of Congress Cataloging-in-Publication Data
Gott, J. Richard.
Sizing up the universe : the cosmos in perspective / J. Richard Gott III and Robert J. Vanderbei.
    p. cm.
Includes index.
ISBN 978-1-4262-0651-1 (hardcover)
1. Astronomy--Charts, diagrams, etc. 2. Solar system--Charts, diagrams, etc. I. Vanderbei, Robert J. II. Title.
QB65.G68 2010
523--dc22

                        2010024503

Printed in China

10/RRDS/1